PLACE AND THE POLITICS OF IDENTITY

If Enlightenment Man stands discredited as gendered and ethnocentric, then the individual has become a contested category. Injustices and old polarities, naturalized in our imagined geographies where everybody is ascribed a proper place, can be deconstructed through analyses of the politics of location.

Through the actions of the new social movements, new political subjects have been created, asserting the unfixed character of identity. A cultural politics of resistance, exemplified by black politics, feminism and gay liberation, has developed struggles to turn sites of oppression and discrimination into spaces of resistance. This new cultural politics employs a richly spatialized vocabulary, focusing on how identity is forged.

Place and the Politics of Identity explores the politics of place and confronts the place of politics. Bringing together some of the most radical voices, both new and established, it challenges received notions of geography and opens a dialogue between cultural politics and geography. Presenting a wide range of detailed case studies, the authors raise new notions of space, politics and identity and explore the interface between theories of cultural difference and everyday realities of political practice.

PLACE AND THE POLITICS OF IDENTITY

Edited by Michael Keith and Steve Pile

Routledge
Taylor & Francis Group
LONDON AND NEW YORK

First published 1993
by Routledge
2 Park Square, Milton Park, Abingdon, Oxfordshire OX14 4RN

Simultaneously published in the USA and Canada
by Routledge
711 Third Avenue, New York, NY 10017

First issued in hardback 2016

Routledge is an imprint of the Taylor and Francis Group, an informa business

Typeset in Baskerville by
Pat and Anne Murphy, Highcliffe-on-Sea, Dorset

British Library Cataloguing in Publication Data
A catalogue record for this book is available from
the British Library
ISBN 0–415–09008–3 ISBN 0–415–09009–1 (Pb)

Library of Congress Cataloguing in Publication Data
Place and the politics of identity/edited by Michael Keith
and Steve Pile.
p. cm.
Includes bibliographical references and index.
ISBN 0–415–09008–3
1. Identity. 2. Discrimination. 3. Human geography.
I. Keith, Michael, 1960– . II. Pile, Steve, 1961–
HM291.P495 1993 304.2–dc20 93-9979 CIP

ISBN 13: 978-1-138-14004-2 (hbk)
ISBN 13: 978-0-415-09009-4 (pbk)

CONTENTS

Contributors vii

1 INTRODUCTION PART 1: THE POLITICS OF PLACE 1
Michael Keith and Steve Pile

2 INTRODUCTION PART 2: THE PLACE OF POLITICS 22
Michael Keith and Steve Pile

3 CLASS RELATIONS, SOCIAL JUSTICE AND THE POLITICS OF DIFFERENCE 41
David Harvey

4 GROUNDING METAPHOR: TOWARDS A SPATIALIZED POLITICS 67
Neil Smith and Cindi Katz

5 LOCATING IDENTITY POLITICS 84
Liz Bondi

6 WOMEN'S PLACE/EL LUGAR DE MUJERES: LATIN AMERICA AND THE POLITICS OF GENDER IDENTITY 102
Sarah A. Radcliffe

7 READING *ROSEHILL*: COMMUNITY, IDENTITY AND INNER-CITY DERBY 117
George Revill

8 POLITICS AND SPACE/TIME 141
Doreen Massey

9 BLACK TO FRONT AND BLACK AGAIN: RACIALIZATION THROUGH CONTESTED TIMES AND SPACES 162
Barnor Hesse

10 THE SPACES THAT DIFFERENCE MAKES: SOME NOTES ON THE GEOGRAPHICAL MARGINS OF THE NEW CULTURAL POLITICS 183
Edward Soja and Barbara Hooper

11 QUANTUM PHILOSOPHY, IMPOSSIBLE GEOGRAPHIES AND A FEW SMALL POINTS ABOUT LIFE, LIBERTY AND THE PURSUIT OF SEX (ALL IN THE NAME OF DEMOCRACY) 206
Sue Golding

12 CONCLUSION: TOWARDS NEW RADICAL GEOGRAPHIES 220
Michael Keith and Steve Pile

Index 227

CONTRIBUTORS

Liz Bondi is a lecturer in Geography at Edinburgh University. She has researched widely around issues of gender and gentrification and feminist geography and is the co-editor of *Education and Society: Studies in the Politics, Sociology and Geography of Education* (1988) and the author of a forthcoming book on feminist geography.

Sue Golding works at the Centre of Theoretical Studies at the University of Essex. She has written widely on the concept of democracy and is the author of *Gramsci's Democratic Theory: Contributions to a Post-Liberal Democracy* (1992). She is a long-time activist in gay and feminist politics, particularly in struggles for sexual freedom.

David Harvey is the outgoing Halford MacKinder Chair of Geography at the University of Oxford. He is a leading scholar of contemporary Marxism and his previous books include *Social Justice and the City* (1973), *The Limits to Capital* (1982), *The Urban Experience* (1989) and *The Condition of Postmodernity* (1989).

Barnor Hesse is a part-time associate lecturer at the Department of Sociology, University of East London, and is currently engaged in doctoral research on theories of Black politics and diasporic identifications at the Department of Government, University of Essex. He is also co-author of *Beneath the Surface: Racial Harassment* (1992).

Barbara Hooper is a postgraduate who works at the Graduate School of Architecture and Urban Planning, UCLA. She has written a critical analysis of the philosophical and political sources of modern planning doctrine.

Cindi Katz teaches in environmental psychology, cultural studies and women's studies at the Graduate School and University Center of the City University of New York. Her work concerns the production space, place and nature in everyday life. She is the co-editor of *Full Circles: Geographies of Women Over the Life Course* (1993).

Michael Keith works at Goldsmiths College, University of London. His most recent research has focused on the racialization of representations of the urban. He is the author of *Lore and Disorder in a Multi-Racist Society* (1993) and the co-editor of *Hollow Promises: Rhetoric and Reality in the Inner City* (1991) and *Racism, the City and the State* (1992).

Doreen Massey is Professor of Geography at the Open University. She is the author of several books on economic geography, including *Spatial Divisions of Labour* (1984). Her most recent publication (with Paul Quintas and David Wield) is *High-Tech Fantasies: Science Parks in Society, Science and Space* (1992).

Sarah Radcliffe is a geography lecturer at Royal Holloway, University of London. After having done research in Peru on rural migration and peasant households, she is now working on political mobilization, the emergence of female peasant unionists and on issues of identity, 'race' and nationalism. She is the co-editor of *'Viva' Women and Popular Protest in Latin America* (1993) and the author of *Confederations of Gender* (forthcoming).

Steve Pile is senior lecturer in Geography at Middlesex University. He is the author of *The Private Farmer* (1990). He is currently working on issues relating to subjectivity, society and space, with particular reference to popular protest in London's Docklands.

George Revill is a lecturer in Geography at Oxford Brookes University. He has worked on identity paternalism, community and corporate culture in the late nineteenth century.

Neil Smith is Chair of the Department of Geography at Rutgers University. His work involves geography, social theory and historical studies of space and nature. He is the co-editor of *Gentrification of the City* (1986) and the author of *Uneven Development* (1984) and he is completing a study of Isaiah Bowman entitled *The Geographical Pivot of History*.

Edward Soja is Professor of Urban and Regional Planning and Associate Dean at the Graduate School of Architecture and Urban Planning, UCLA. He is the author of *Postmodern Geographies* (1989) and has written widely about the place of space in contemporary social theory, notions of spatiality and theorizations of postmodernity.

1

INTRODUCTION PART 1

The politics of place . . .

Michael Keith and Steve Pile

WHY THINK ABOUT PLACE, POLITICS AND IDENTITY?

Writing in the wake of yet another Conservative party victory in the 1992 British general election and more urban conflict in Los Angeles and various British cities, it would appear that politics has changed very little – no matter where you look – whether at the level of the nation-state or on the streets.

On the other hand, many social commentators have detected some kind of a sea change – not only in the nature of Western societies but also in global economic relations. This transformation is usually referred to as the transition from modernity to postmodernity. Unfortunately, debates on the nature of the modern and the postmodern have become internal and arcane (Pile and Rose 1992). There is no broad agreement on what modernity was, on what post-modernity is, or on how we got from there to here, or on where we are going next.

Much of this debate revolves around three related issues: the relationship between time and space; the potential of politics; and the construction of identity. This book is, however, not about the debate over modernism and postmodernism, instead the question is this: can concrete geographical and historical circumstances – whether the British general election or civil disturbances on the streets of Los Angeles – be understood as expressions of abstract social relations? Many theorists allege that the contemporary (whether it be modernity or postmodernity) is now much more complex than hitherto and that it is no longer good enough to theorize power as the expression of a singular dimension of oppression, such as class or gender or race. We may question how far this is true, but the effort now and here is to specify the relations between the many dimensions of oppression – including class, gender and race – and then to suggest strategies of resistance. In order to articulate an understanding of the multiplicity and flexibility of relations of domination, a whole range of spatial metaphors are commonly being used: position, location, situation, mapping; geometrics of domination, centre–margin, open–closed, inside–outside, global–local; liminal space, third space, not-space, impossible space; the city. Such terms are used to imply a complexity which is never directly explored or confronted (to use two more spatial metaphors) – partly because it is rarely clear whether the space invoked is 'real', 'imaginary', 'symbolic', a 'metaphor-concept' or some relationship between them or something else entirely. For example, the metaphor-concepts of exploration (which has deep roots in imperialism) and confrontation (which implies a face-to-face, potentially violent opposition) may evoke social relationships between authors, texts and readers which are not intended or are inappropriate.

In this introduction, we wish to achieve two things: first, in Part 1, we want to explore the politics of place, and, second, in Part 2, we want to confront the place of politics. In Part 1, we introduce the notion of *spatiality* by drawing on the writings of Fredric Jameson and Ed Soja. This review suggests that space cannot be dealt with as if it were merely a passive, abstract arena on which things happen. The spatialities of urban regeneration and the politics of diaspora illustrate precisely these themes. For the purposes of our argument, the first may be understood as an identity politics of place and the second as the spatialized politics of identity. In Part 2, we ground this discussion in debates surrounding the sense of space evoked by political theorists. At the end of Part 2, we describe how the essays collected here are both embedded in this tradition, and confront it, in analyses of the relationship between place, politics and identity.

THE SPATIAL VOGUE

One of the most prominent commentators on the condition of the contemporary, Fredric Jameson (1984; 1991), has suggested that these new patterns are distinguishable from old ones by the domination of social and cultural life by the logic of spatial organization, rather than time.

> I think that it is at least empirically arguable that our daily life, our psychic experience, our cultural languages, are today dominated by categories of space rather than categories of time, as in the preceding period of high modernism.
>
> (Jameson 1991: 16)

Jameson suggests that there are three basic phases in the development of the spatial logic of society under capitalism. In the first stage, he argues that *market capitalism* was dominated by the spatial logic of the grid. Capitalism organized, and was organized by, a geometrical view of space. This view was subsequently replaced by the growing contradiction between lived experience and structure. In the second stage, *monopoly capitalism*, figurative space stands in the place of absent causes. Space represents, and is represented by, distorted images of the real determinations of social relations. Currently, the spatial logic of *multinational (postmodern) capitalism* is simultaneously homogeneous and fragmented – a kind of 'schizo-space'.

Indeed, for Jameson, schizophrenia seems to have become the mark of the age: old loyalties of class or gender or race fragment, dislocate, rupture, disperse; new loyalties of class and gender and race interrupt, disrupt, recombine, fuse. No one is quite sure of the ground on which they stand, which direction they are facing, or where they are going. Under these circumstances, the subject is proclaimed dead; the agent of history no more.

> But what is involved here is in reality practical politics: since the crisis of socialist internationalism and the enormous strategic and tactical difficulties of coordinating local and grassroots of neighborhood political actions with national or international ones, such urgent political dilemmas are all immediately functions of the enormously complex new international space in question.
>
> (Jameson 1991: 413)

In order to counteract the political paralysis of today, Jameson develops an alternative view of space and political action, provisionally naming it as the aesthetic of cognitive mapping. Jameson is not calling for the mapping of old notions of space, instead this is the name of a new form of radical political culture; its fundamental object is the 'world space of multinational capital' (Jameson 1991: 54). Cognitive mapping is in some senses recognized to be both unimaginable and impossible; it attempts to steer between Scylla and Charybdis, between an awareness of global processes and the inability to grasp totality. Nevertheless, it is also meant to allow people to become aware of their own position in the world, and to give people the resources to resist and make their own history. It is the logic of capital itself which produces an uneven development of space. These spaces need to be 'mapped', so that they can be used by oppositional cultures and new social movements against the interests of capital as sites of resistance.

The problem for oppositional politics is that 'everyone "represents" several groups all at once' (Jameson 1991: 322). This means that the identity of the subject position and of the political movement need to be understood simultaneously. This brings us back to what political pundits in Britain have described as 'Basildon man' and 'Essex girl'. These are geographically specific, gendered stereotypes of working-class people who vote against their class interest by voting for the Conservative party – the party of capital. We should not forget, in the nostalgia for the simplicity of class war, that its rhetoric is increasingly useless – because, however fleetingly, it does not work. Identities supplant others – no matter how important an 'objective' circumstance or central an identity is argued to be by politicians, academics and pundits. Following Lefebvre, Jameson argues that what is needed, in order to help recover the sites of resistance, is

> a new kind of spatial imagination capable of confronting the past in a new way and reading its less tangible secrets off the template of its

> spatial structures – body, cosmos, city, as all those marked the more intangible organization of cultural and libidinal economies and linguistic forms.
>
> (Jameson 1991: 364–5)

This contention suggests, at least, that space may be the template from which the secrets of reality are to be read. Later, Jameson notes – on the last page of his book – that work in this vein may best be exemplified by the writings of Ed Soja (Jameson 1991: 418).

However, Soja, similarly drawing on Lefebvre but also more on Foucault, does not see space as so passive, undialectical. Both Soja and Jameson share a common concern for spatiality, partly because this term is designed to reinstate space at the heart of a dynamic conception of time–space relations. But Soja wants to locate his argument on different terrain from Jameson; while Jameson sees space as a process of distance, Soja would rather treat distance as a dialectic between separation and the desire to be close. This leaves the question of the individual's occupation of subject positions in a different conceptual place. For Jameson, the individual is to be mapped by the spatial specificity of their subject positions, in order to uncover the hidden human geography of power, but Soja's schema suggests that even this dynamic understanding of the situation is too solid: space is not an innocent backdrop to position, it is itself filled with politics and ideology.

> We must be insistently aware of how space can be made to hide consequences from us, how relations of power and discipline are inscribed into the apparently innocent spatiality of social life, how human geographies become filled with politics and ideology.
>
> (Soja 1989: 6)

Soja (1989: 7, 122–6) argues that space has been misrecognized by contemporary social theory. It has suffered from a dual illusion: either space has been seen as opaque or as transparent. The *illusion of opaqueness* has led to a concentration on concrete forms, where space is fixed, dead and undialectical (following Foucault). What is lost from view are 'the deeper social origins of spatiality, its problematic production and reproduction, its contextualization of politics, power and ideology' (Soja 1989: 124). The *illusion of transparency* dematerializes space, it becomes an abstraction, a supposedly real representation of concrete forms: 'spatiality is reduced to a mental construct alone . . . Social space folds into mental space . . . [and] away from materialized social realities' (Soja 1989: 125). This version echoes Jameson's identification of geometrical space, but connects the representation of space to actual space. Having made this connection, Soja is able to argue that the contemporary situation is marked by the convergence of three different kinds of spatialization: posthistoricism, post-Fordism and postmodernism. These, Soja continues, may now be reconnected – in a mutually reinforcing hermeneutic arc – to Jameson's radical cultural politics: i.e. cognitive mapping. It is the mapping of these features of space which will allow 'a new way of seeing through the gratuitous veils of both reactionary postmodernism and late modern historicism to encourage the creation of a politicised spatial consciousness and a radical spatial praxis' (Soja 1989: 75).

If, as Soja argues, 'the geography and history of capitalism intersect in a complex social process which creates a constantly evolving historical sequence of spatialities' (Soja 1989: 127), then certain questions are invoked – and these relate to the way in which place, politics and identity are to be understood through an already spatialized array of concepts, such as mapping and spatiality. These issues may be introduced by turning to bell hooks, who is fast becoming a shibboleth for white academic men – including us – who want to prove beyond any shadow of a doubt their radical credentials.

QUESTIONS FOR MAPPING THE POLITICS OF IDENTITY

> Our living depends on our ability to conceptualize alternatives, often impoverished. Theorizing about this experience aesthetically, critically is an agenda for radical cultural practice. For me this space of radical openness is a margin – a profound edge. Locating oneself there is difficult yet necessary. It is not a 'safe' place. One is always at risk. One needs a community of resistance.
>
> (hooks 1991: 149)

Distinct, irreconcilable understandings of space underscore the cultural mappings of the contemporary. For Jameson, space is a template, while for Soja, such a geometrical conception of space is passive, fixed, undialectical and no longer appropriate. For hooks, both these perspectives involve risks and dangers which are directly political; for those who have no place that can be safely called home, there must be a struggle for a place to be. Her evocation of the margins is simultaneously real and metaphorical – it defines an alternative spatiality: radical openness. A different sense of place is being theorized, no longer passive, no longer fixed, no longer undialectical – because disruptive features interrupt any tendency to see once more open space as the passive receptacle for any social process that cares to fill it – but, still, in a very real sense about location and locatedness.

In this collection, the authors develop their own lines of disruption; new spaces of politics are identified, new politics of identity are located. Unwisely, perhaps, we would like to suggest that there are three key areas that distinguish these new projects:

1 locations of struggle;
2 communities of resistance;
3 political spaces.

These surfaces of articulation permit alternative agendas for geography and for those interested in space and place for other reasons. Key strategic moves are being made by the authors in this book, ones which transgress and displace traditional notions of space and place, of time and history. New spaces of resistance are being opened up, where our 'place' (in all its meanings) is considered fundamentally important to our perspective, our location in the world, and our right and ability to challenge dominant discourses of power.

> As a radical standpoint, perspective, position, 'the politics of location' necessarily calls those of us who would participate in the formation of counter-hegemonic cultural practice to identify the spaces where we begin the process of re-vision.
>
> (hooks 1991: 145)

It is in this spirit of re-visioning that this volume charts attempts to de-limit positions which avoid two equally unacceptable arguments: a myth of spatial immanence and a fallacy of spatial relativism. The first is the notion, self-evidently bizarre on close inspection, alarmingly common in much social description, that there is a *singular*, true reading of any specific landscape involved in the mediation of identity. On the other hand, it is invidious and disingenuous to suggest that each and every reading of a specific landscape is either of equal value or of equal validity; such notions lead to an entirely relativist notion of spatiality.

Instead, it may be argued that simultaneously present in any landscape are multiple enunciations of distinct forms of space – and these may be reconnected to the process of re-visioning and remembering the spatialities of counter-hegemonic cultural practices. We may now use the term 'spatiality' to capture the ways in which the social and spatial are inextricably realized one in the other; to conjure up the circumstances in which society and space are simultaneously realized by thinking, feeling, doing individuals and also to conjure up the many different conditions in which such realizations are experienced by thinking, feeling, doing subjects.

The spatialities in which we are interested in this volume are the source of both the complexity of our understandings of the spatial and the confusion in the contemporary vogue for a spatialized vocabulary. Most readily seen in the unproblematic use of metaphors of, and allusions to, the spatial, there is a sense in which the geographical is being used to provide a secure grounding in the increasingly uncertain world of social and cultural theory. As some of the age-old core terms of sociology begin to lose themselves in a world of free-floating signification, there is a seductive desire to return to some vestige of certainty via an aestheticized vocabulary of *tying down* elusive concepts, *mapping* our uncertainties, and looking for *common ground*.

Transparent landscapes: the myth of spatial immanence

At times, the resort to spatialized vocabulary deploys limited and misleadingly unproblematic evocations of spatiality. Typically, in an acknowledgement of the increased salience of notions of the spatial to contemporary social theory, Scott Lash and Jonathan Friedman (1992) have contrasted the fixity of pre-modern identities (founded in, for example, religion) with the manner in which 'social space opens up the way for the autonomous definition of identity' in modernity (Lash and Friedman 1992: 5). They also draw attention to the importance of spatial scale: pointing out that Marshall Berman's (1988) analysis of modernity begins with the localism of place, and citing Jane Jacobs' (1961) concern with the neighbourhood in the life and death of cities. Yet in spite of this welcome acknowledgement, they then go on to draw a tendentious equivalence, respectively, between notions of the public and the private, the universal and the local, and landscape and vernacular spaces (Lash and Friedman 1992: 19).

The latter dichotomy – taken from the cultural geography of J. B. Jackson – has been drawn on even more explicitly by Sharon Zukin (1991; 1992) in describing the manner in which landscapes of power may triumph over the vernacular (1991, Chapter 9). In her analysis of the cultural forms of contemporary capitalism, Zukin tends towards economic reductionism in an opposition between '*markets* – the economic forces that detach people from established social institutions – and *place* – the spatial forms that anchor them to the social world, providing the basis of a stable identity' (Zukin 1992: 223). In order to analyse this transition, Zukin introduces the notion of 'liminal spaces'. She argues that the localism, or neighbourhood urbanism, of the modern city has been transformed into postmodern transitional space. This space is 'betwixt and between' economic institutions but is best described by the adjective liminal because it 'complicates the effort to construct spatial identity' (ibid.: 222). Liminal spaces are ambiguous and ambivalent, they slip between global markets and local place, between public use and private value, between work and home, between commerce and culture. Nevertheless, she time and again returns to an evocation of a singular immanent meaning which lies buried beneath the surface and awaits revelation. A scene is either 'landscape' or 'vernacular', a one-to-one correspondence between the image and consumption of a place and its *reality*. The implication here is that multiple meanings (liminal landscapes) reflect only the multi-faceted images of capital rather than other sources of multiplicity (ibid.: 240).

Zukin provides an example of what she is talking about: a tour of the streets of Clerkenwell. The tour leader – who is a teacher and a recent in-migrant – guides the few visitors around the area, giving them a view of the cityscape which allows the city to be consumed as an ensemble. The city is subjugated to no particular narrative of history; the historic parts were merely presented as cultural artefacts to the audience. This is a liminal scene; the city is created as a cultural category. For Zukin, this tour exemplifies the way in which the sense of place has succumbed to market forces. Thus, the postmodern urban landscape imposes multiple perspectives which are not only wedded to economic power but also facilitate 'the erosion of locality – the erosion of the archetypal place-based community by market forces' (Zukin 1992: 240).

Our point here is not to dismiss Zukin's work under the abusive rubric of economism, but instead to suggest that this kind of economic base-cultural superstructure analysis taps just one form of spatiality which structures landscape, but never exhausts its meaning. Plagiarizing desperately, landscape is made in the image of capital but this is not its sole image. When we walk the streets – whether as guides, tourists, or inhabitants – we are at once invoking a host of competing spatialities, not a straightforward spatialized reference with a correspondent true meaning.

Anything goes? The perils of spatial relativism

The search for order, the sense of digging behind the cultural façade to find the one true meaning of landscape is sometimes, explicitly or implicitly, prompted by a fear that if every individual reading of geographical form is equally true, then geographical analysis falls prey to cultural relativism – in other words, that *anything goes*. It is in this context that the work of Walter Benjamin can provide a useful counterpoint here; certainly his street tour is a long way from Zukin's.

In capturing the cities of Berlin and Paris in his writing, Benjamin creates, through disparate fragments of prose, a world that on first appearance is exclusively his own. And in aestheticizing the spaces of the city, he seems to go even further, turning prose into poetry. Well, yes and no. Because Benjamin does much more. As the exemplary *flâneur*, his work defines a sensibility that takes the specific to rise beyond the particular, prefiguring the urban readings of Barthes and de Certeau and the politics of situationism, he finds the universal through a street plan, humanity through his arcades.

But Benjamin's arcade is in no way a grid reference or a mere location. It is a way of life, a metaphoric allusion to a form of sensibility, a Proustian metonym, an invocation of a way of seeing, a nodal point in a field of vision that condenses sets of contradictory meanings. It is all of these things and more. And none of them is identical. Each is closely related to most of the others but each evokes a slightly different form of spatiality. As Susan Sontag has commented on Benjamin's work *One Way Street*, 'reminiscences of self are reminiscences of a place, and how he positions himself in it, navigates around it' (Sontag 1979: 19).

We cannot read the street straightforwardly. Famously, 'botanizing the asphalt' as *flâneur* immersed in the urban experience; the whole work of Benjamin is about the intertwining of experience, knowledge and spatiality. Visible or invisible, Benjamin knows the city and the street through a vision that may not be corporeal and may not acknowledge the gendered character of his own gaze (see Pollock 1988) but most certainly celebrates the conflation of the sites of the urban and the sight of the city. Places are known through this sensibility, but places also, in turn, constitute the sentient individual.

As Sontag at one point suggests, the whole opus of Benjamin might be called *A la recherche des éspaces perdues*: 'Benjamin is not trying to recover his past, but to understand it: to condense it into its spatial forms, its premonitory structures' (Sontag 1979: 13). In this sense, Benjamin's work provides us with an excellent illustration of the complex relationship between spatialities and identity which forms our focus of interest in this volume. Space too is ambiguous, ambivalent, multi-faceted, duplicitous (Daniels 1989). Difficult.

In part, reference to Benjamin is a useful reminder of the strictly licensed 'novelty' of a spatialized vocabulary of social theory. But, more significantly, his work throws light on the relation between identity and the spaces through which identity is both produced and expressed. The spaces of Benjamin's Paris and Berlin are both real and metaphorical simultaneously. They are not just a personal view but then they are not the true representation of city society either. Too often used as a residual descriptive container which defines the empirical, these spatialities are instead to be understood as a constitutive element of the social. Neither are these spaces ethically rudderless. Politically, it should not be forgotten that Benjamin's work provides the literary provenance for Theodor Adorno's powerful analysis of the interaction of the epistemological and the aesthetic.

We are suggesting a more complex relationship between the so-called real and the so-called metaphorical: one does not merely cover the other; one is not more real than the other. For example, to return to the work of Zukin and her description of Disneyland, where

> stage-sets evoke the social production of visual consumption, with its history of resort and fantasy architecture, its fictive nexus in Disney World, and its dependence on markets to foster products that in turn create a sense of place. In this landscape, socio-spatial identity is derived purely from what we consume.
>
> (Zukin 1992: 243)

This bleak formulation – I consume therefore I am – functions not as critique (as it does in the artwork of Barbara Kruger's formula 'I shop therefore I am') but as a normative description.

Her analysis implies, demands, the recovery of authentic, good landscapes, which contrast to the Mickey Mouse worlds of capital. Such nostalgia is unreal: how can the authentic be authenticated – or, more properly, who is to authenticate the vernacular? We would argue that spatialities draw on a relationship between the real, the imaginary and the symbolic that is not beyond truth and falsity, but is different from it. Is Hardy's Wessex a reliable source any more or less than Rushdie's India? Well, yes and no. It depends.

We can develop this argument about the 'dependency of truth claims' further through a relatively prosaic example. A few years ago, certain Labour party-controlled local authorities in the UK, particularly those in London, chose to describe themselves as 'nuclear-free zones'. It is impossible to evaluate such designations purely in terms of the putative truth or falsity of their meaning. Assessed in the spirit of literalism, such designations were always manifestly absurd – given the failure of the contemporary nuclear device to respect the borders of the territorially inclined bureaucrat.

But the notion of nuclear-free zones was never intended to stand for such literal interpretation. They were meant to evoke a particular kind of politics by an appeal to the emotions that people felt about the places where they lived, their communities, their localities – their homes. Nuclear-free zones were intended to make links: they stood as territorial metaphors of a particular kind of 1980s politics – a different politics of consumption that tried to unite the many different fractions of consumption and production classes behind a progressive political platform. Not so much 'I consume therefore I am' as 'I die therefore I am'.

So how should the nuclear-free zone be judged? As true or false? As real or metaphorical? As authentic or unauthentic? As true as a burning breast or as false as a bleeding heart? Quite clearly they represented a particular invocation of spatiality. And, given the course of events in the late 1980s, it might be argued that nuclear-free zones were cursed by English anti-intellectualism and Anglo-Saxon empiricism – the preferred politics of the literal-minded – which hijacked the symbolism and stranded nuclear-free zones in civic debates about the price of signposts and the cost of rerouting the trains carrying nuclear waste.

Much can be said for and about the expenditure priorities and more immediate health hazards, but the Right used this particular tactic to lampoon the Left as 'loony', and ultimately was yet another nail in the coffin of local democracy. The nuclear-free zones were literally daft, but the strategy – right or wrong, good or bad – worked by deploying a spatialized, political language. It created, however, briefly, a new space of resistance that tried to weld place, politics and identity, but succeeded only in provoking a backlash from a Thatcher government tired of thorns in its side.

The question we would like to ask is this: if nuclear-free zones were so daft, why did they require a response at all? We would like to suggest two dimensions of an answer, based on one premise. The premise is that the tactic could not be ignored because the nuclear-free zones foregrounded something that was meant to remain on the margins of public debate – the stupidity of a defence policy based on Mutually Assured Destruction (i.e. MAD) – at the heart of the public arena of national politics. This presentation of the margins at the political centre contains two basic strategies: the identity politics of place and the spatialized politics of identity. Neither strategy carries a guarantee of success but may be illustrated by two examples: one drawn from the politics of urban regeneration in London's Docklands and the second from the politics of the diaspora.

'THIS LAND IS OUR LAND': TERRITORIALIZED POLITICS IN LONDON'S DOCKLANDS

The recent spectacular crash of Olympia & York in May 1992 has put the planned £3 billion development of Canary Wharf in question and focused national and international attention on London's Docklands. On Guy Fawkes night 1992, a satirical TV programme, *Alas Smith and Jones*, found a funny side. Drawing on a scene from the musical version of Charles Dickens' *Oliver*, the players sang: 'Who will buy?' Good question: who will buy the Canary Wharf development? There are levels of irony here. The film *Oliver* sanitizes the London that Dickens once walked through. The scene in the movie was shot in one of Bath's most prestigious crescents – the singers and dancers have clearly never missed a hot meal or a hot bath in their lives. Canary Wharf is an island of wealth: red and green marble has been brought from Italy and Guatemala; young trees brought from Canada; and the quality of the development is of the highest-order (Punter 1992). The Tower (1 Canada Place) – now so much the symbol of Docklands, but originally intended to be only one of three skyscrapers – stands 'proud' amidst some of the most deprived estates in one of the most deprived boroughs not just in London but in the country. This disparity is resented.

Even during the building of the project, its financial prudence and its appropriateness were being questioned. A local doctor remembers that, during the topping-out ceremony of the 850-foot Tower, one bank manager put it this way: 'fucking white elephant if you ask me' (Widgery 1991: 159). This caustic observation must be understood against several contexts: first, the political economy of the redevelopment of London's docks; second, the impending financial collapse of Olympia & York (at the time of the remark still only predicted by the most pessimistic of city analysts and optimistic of leftist radicals); and, third, the feelings and actions of the people who live in the shadow of the Tower – physically and metaphorically. It is this last aspect with which this subsection is most concerned, because the response of local people to the proposed redevelopment of the docks and the imposition of the London Docklands Development Corporation can be seen a particular political mobilization of place and identity.

A dog's breakfast: the political economy of regenerating the docks

In the 1950s and 1960s, despite the reconstruction of London's docks after the Second World War, it became clear that they were becoming uneconomic. Several factors contributed to this situation: partly as a result of the war, London's role as a major trade centre had been undermined; there was an overcapacity on upstream docks and warehouse facilities; manufacturing firms used war damage compensation to move out of the East End and into the new towns being built outside London; much of the transport system had fallen derelict; and, many buildings had been destroyed by the Blitz and those that were left had become obsolete. The two most important factors, however, appear to have been: first, that changes in maritime technology required larger, deeper docks with the capacity to take container ships (Ambrose 1986: 218); and, second, that the Port of London Authority realized that the land around the docks would be more profitable if it were used for office expansion and if the docks' functions were relocated to Tilbury. The combined result was that there was ever-decreasing investment in the docks and the East End more generally. In 1967, the East India Dock became the first to close and the rest of the docks were quick to follow, until the Royal Docks ceased operation in 1982 (Hardy 1983; Brownill 1990: 19; Coupland 1992).

By the early 1970s, London's docks and the land around them had been identified as the largest opportunity for redevelopment in Europe (see Ambrose 1986: 221; Brownill 1990: 1). The closure of the docks meant that 22 square kilometres of land had become redundant – the problem was what to do with all that space. Unfortunately, there was a wide range of interest groups which each had their own answers to this question, including a sizeable local population with strongly assertive political traditions (see Ambrose 1986). This produced a situation of bewildering complexity. Throughout the 1970s, a stream of reports and alternative plans emerged from a succession of agencies; these documents highlighted sharp divisions of opinion. For example, two main issues were identified: employment and housing. On employment, one side argued that any kind of development – including office blocks, leisure, the service sector and alike – should be allowed if it meant jobs; while the other side argued that new employment should be matched to the skills of the people who already lived in Docklands. On housing, some people wanted private housing speculation which would create a 'normal' tenure balance; on the other hand, people wanted more low-rent public housing. The problem was further complicated because some people wanted homes rather than offices. Centrally, the community groups wanted the power to be able to control development, and to control the resources to enable development.

Even so, different interest groups took different views, and these would alter in particular situations and over time. Broadly, four arenas of disagreement can be identified, mutually opposing views were held not only within each arena, but also between groups in each arena. There were conflicts between:

1 commercial, industrial and community interests;
2 the Greater London Council, each of the five local boroughs (Tower Hamlets, Newham, Southwark, Lewisham and Greenwich) and the Port of London Authority;

3 local government and central government;
4 various representatives (from local councillors through the chairs of local community groups to members of other groups, like Docklands Childcare Campaign) of 'the community' and people resident in and around Docklands.

By the end of the 1970s, the development of London's Docklands had become a Gordian knot. In 1979, the then newly elected Conservative government cut that knot. Within months of being elected, Michael Heseltine (then Secretary of State for the Environment) was proposing to set up Urban Development Corporations (UDCs), with centrally appointed boards and access to state funding. One was to be set up in London's Docklands.

These UDCs were based on the assumption that only private investment, and not state planning, could save Docklands; so, the purpose of the UDCs was to do everything possible to get private and institutional money into the area, while simultaneously preventing public intervention in the redevelopment of the area. The UDCs were enabled by the powers of the 1980 Local Government, Planning and Land Act; which also relaxed planning controls, enabled the compulsory sale of public land, set up inducements for private capital, and controlled public spending.

In July 1981, the London Docklands Development Corporation (LDDC) was set up under the chairmanship of Nigel Broakes. For Nigel Broakes, regeneration was not about developing Docklands in ways that suited the needs of local people, and he was certainly not interested in allowing the participation of local people in decision-making. In the initial period, local people were denied access to the decision-makers who were about to alter their lives radically.

Despite being a supposedly *laissez-faire* free market, huge sums of money have been channelled into the LDDC. By 1991, £1,134 million had been spent by the LDDC alone. Unsurprisingly, a lot has happened in Docklands. By March 1989, there had been £6 billion of private sector investment; 17,000 dwellings had been built or started (i.e. about three a day since the inception of the LDDC); 20,000 jobs had been created; and, 0.81 million square metres of floor space had been built (see Brownill 1991: 1). Canary Wharf alone is massive: the access route, Westferry Circus, is the size of Trafalgar Square in central London. It will have 0.95 million square metres of floor space, which is of equivalent size to Manchester's. It will consume the same amount of electricity as a city the size of Oxford, and steel used in the construction will amount to one-third of the United Kingdom's annual output. Altogether the LDDC estimated that there will be 30,000 dwellings built, 200,000 jobs created, and 1.9 million square metres of office space.

Nevertheless, it can easily be demonstrated that regeneration has not helped local people. Since 1981, homelessness has increased by 304 per cent, while it increased 66 per cent in inner London as a whole. Of the 17,000 dwellings that have been started, 85 per cent are private developments, while only 5 per cent are council houses. Total employment is said to have risen to over 42,000, but this represents a real rise of only 10,000 because of job losses in the area. Moreover, 44 per cent of those working in Docklands travel in from outside. Of the 20,000 supposedly new jobs created, 75 per cent are transfers from elsewhere. It is not surprising, therefore, to find that unemployment levels are twice the rate of the London average (i.e. some 12–13 per cent in the Docklands boroughs, but 7–8 per cent in London as a whole). All this despite over £7 billion investment over the so-called 'decade of achievement'.

The resistances of Docklands' people: different spaces, different politics

From the beginning, there was a high degree of conflict about the development of London's Docklands. People were especially annoyed about the imposition of the LDDC, which meant the removal of local democracy. Local people had no way of representing their views to decision-makers, and the decision-makers did not consult local people.

> By the time the LDDC was set up, there were already a large number of community organisations in existence. . . . These groups had definite demands for how the land should be used in Docklands and for greater community involvement.
>
> (Brownill 1990: 108)

There were (and are) specific geographies to these protests: people organized in different ways at different times in different places. The notion of Docklands became a symbol around which people mobilized; a way in which residents identified their neighbourhood; and an administrative and economic zone; an imagined geography and a spatialized political economy – a way of seeing and a way of life. Nevertheless, it is useful to categorize these groups into three types: Docklands-wide, area specific, and interest based.

1 *Docklands-wide*: such as Joint Docklands Action Group, Docklands Consultative Committee, Docklands Forum;
2 *Area specific*: such as Wapping Parents' Action Group, Association of Island Communities, Association of Wapping Organizations, Newham Docklands Forum, North Southwark Community Development Project, Rotherhithe Community Planning Centre;
3 *Interest based*: such as trades unions, church organizations, and special interest groups.

The intersection between these organizations led to a multi-layered and multidimensional geography of resistance. Partly as a result of the range of groups and their different interests, partly because of the powers given to the LDDC, and partly due to the geographical dislocation between the various Docklands communities, resistance tended to be fragmented and frustrated. Nevertheless, people were tied to their land, and they wanted to have a say. As one campaigner said in 1987: 'There are spaces that we can open up and where we can win things, we are not interested in romantic failure' (Peter Dunn, Docklands Community Poster Project; cited in Nairne 1990: 189). Different spaces of resistance were opened up, and there were some successes.

Across Docklands, an attempt to articulate people's concerns was made through the production of Community Murals by Docklands Community Poster Project (see Evans and Gohl 1986: 86–90; Nairne 1990: 182–9; Wallis 1991: 300–3). In 1984 (pre-dating John Major's Citizen's Charter), community groups produced 'The People's Charter for Docklands' which was taken to Parliament by a 'People's Armada' on 17 April 1984. If Docklands is linked by anything, then it is the river; so in 1985 and 1986 community groups organized more People's Armadas; each containing a flotilla of barges, including one with a banner saying 'Docklands Fights Back', reminding everyone who saw it that Docklands communities had not disappeared.

Other forms of resistance were local and the result of local circumstances. For example, in 1983, Newham Docklands Forum produced a 'People's Plan' for the Royal Docks area in which they outlined their proposals to address poor housing and economic decline. Although this plan was not adopted, there were positive effects – the community gained a Plan Centre, a laundrette, a training centre and a crèche in Pier Parade. This may not seem like much, but this was an extremely run-down area. At King's Cross, where greater planning gains were possible, this strategy also proved effective. And now, in 1993, Newham Docklands Forum are drawing up a second 'People's Plan' in order to fill the planning and development vacuum created by the recession and by the end of the LDDC.

One successful campaign was in the Surrey Docks, where there was a riverside open space at Cherry Gardens. The LDDC identified this as a prime site for luxury homes, but it had been acquired by Southwark Council and they intended to build housing for rent and allow riverside access. Nevertheless, the site was vested by the LDDC, who proposed to build four pairs of seven-storey luxury blocks on the waterfront. Local people were incensed, and a vigorous campaign was mounted, backed by the Rotherhithe Community Planning Centre. The LDDC was left in no doubt as to the strength of local feeling. So, in March 1985, they offered one-third of the site to housing associations for rented housing. Fortuitously, while the argument raged, archaeological remains were found on the site (believed to be Edward III's manor house). This tipped the balance: in January 1986, the LDDC offered half the site back to Southwark Council, who promptly built low-density housing for local people.

Up until 1987, the LDDC was ascendant, obvious changes had occurred in Docklands, and opposition to the LDDC had been successfully excluded from the decision-making process. The Cherry Gardens success was at least partly the result of an archaeological stroke of luck. Nevertheless, problems were already becoming all too visible. In part, these were related to the crash of the stock exchange on 'Black Monday' in October 1987; to misguided spending, especially on the hopelessly inadequate Docklands Light Railway; and to a failure to provide adequate transport links, such as the extension of the Jubilee underground line. These events, and the continued actions of community groups, have forced on to the agenda the other sides of the LDDC's 'Ten Years of Achievement'. Docklands is (in)famous for Canary Wharf, post-modern architecture, and gentrification, but two other Docklands are now seen to exist: the Docklands belonging to the indigenous communities and the Docklands that cannot be sold for love nor money. Under these circumstances, perhaps the greatest achievement of the community groups is that they have not gone away. By mobilizing a territorialized sense of both place and community identity, they forced themselves on to the political agenda and, because of their continued commitment to 'their land' (even though they neither own it nor control it), they will outlast the LDDC and continue to resist the fly-by-night property developers.

DIASPORIC POLITICS AND POST-COLONIAL AUTHORITY

The way in which we talk in everyday language is routinely spatially marked (Keith 1988; Pile 1990). Frequently, in efforts to speak to others, to empathize or to generalize, this marking is toned down, suppressed, even eliminated. Again, we see a particular kind of slippage by which the constitutive spatiality of 'being' is *sotto voce*, muted in a bid for universality. There is a rhetorical root to such a move. Discourses ostentatiously marked with their spatiality are conventionally assumed to be narrow-minded, bounded – coming complete with a self-confessed specificity that, it is frequently assumed, restricts their relevance. Working within common sense understandings of knowledge, the markings of spatiality can become the stigmata of parochialism.

In this sense, narratives of identity formation in mainstream social science have frequently spoken to an interplay of commonality and difference that erases spatiality through a homogenization of the specific – not a process of *misrepresentation* through over-generalization but instead a naturalization of particular experiences within a frequently implicit spatial frame of reference. Typically, this frame of reference operates at a higher level of generalization, such as the silent evocation of (national) societies or the reified subject positions of either post-colonial subjection or Eurocentric domination.

This is, in part, de Certeau's point when he argues that 'normative structures have the status of spatial syntaxes' and that 'every story is a travel story – a spatial practice' (de Certeau 1984: 115). This spatialized syntax is frequently not obvious, even if it is invariably present (if only as an absence). Just as all knowledge on close inspection is both empowered and restrained by its situated generation, all narratives can be unpacked to reveal the frequently implicit spatialities that they evoke, varying from the mundane to the contradictory.

An exemplary case of this is the manner in which popular, media and academic discussion of the experiences of British Black communities frequently implies the generality of a space–time rubric that is loosely aligned with postwar consensus politics and Fordist labour migration. Yet the narratives of Black communities in the port cities of London, Bristol and Liverpool, predating Fordist labour demand, have their own historicity and spatiality that may have marked the consciousness, practice and racialization of migrant communities, though you might never think so from reading most accounts of 'Black Britain'. It is precisely these traces that Barnor Hesse is interested in examining in Chapter 9 of this volume. The distinct experiences of different migrant communities with African and Caribbean origins is inscribed in the production of space at the local level, and yet nationally such communities are bracketed by a single *articulating principle* that captures British Black experiences in the vocabulary of the postwar settlement.

Likewise, particular realizations of colonial ideology may have been inscribed in locally specific ideologies of empire; these underscored the definitions of 'self' and 'other' that lay at the heart of spatially diverse and contradictory understandings of nation, whiteness, power, subjection, Commonwealth; and which were installed at the heart of the imperial metropolis (C. Hall 1992).

In contrast to this sort of silenced spatiality, the diversity of different migrant histories commonly now draws on the geographically rich notion of diasporic politics. In this context, the more recent work of Paul Gilroy provides a useful illustration of a project to reconcile precisely the sort of dilemmas we have been working through here by using the imagined geographies of the Black diaspora. Gilroy has formulated what can be read as a strategic double-take on the recurrent tensions of a politics of identity. On the one hand, he is attempting to avoid the ethnic essentialism of some forms of identity politics whilst claiming that racialized differences remain significant and at times incommensurable.

The corollary of this is an acknowledgement that, whilst the critiques of Enlightenment reason associated with postmodernity have a power that needs to be recognized, he does not want to let go of the gains that particular forms of rationality offer in fighting racialized oppression. Gilroy sometimes expresses this in terms of an outline of the potential for a specifically Black modernism. Loosely defined, his project appears to work on narrative and conceptual levels, with the notion of an imagined spatiality of diasporic politics serving to mediate these tensions. At the level of historical narrative, the project stresses the international links between Black intellectuals throughout the last hundred years and more; Black diasporic intellectual forms intermingled both within a diasporic international context and with Western thought as well. Black nationalism is tied to Hegelian thought, just as Pan-Africanism resonates in parallel struggles across and beyond the diaspora.

But more significantly for our argument, the diaspora invokes an imagined geography, a spatiality that draws on connections across oceans and continents and yet unifies the Black experience inside a shared *territory*. This experience is the source of difference and yet does not legitimate the elevation of 'the Black experience' to an incommunicable cultural essence. It is a spatialization of Black consciousness which is not controlled by those who 'would police black cultural expression in the name of their own particular history or priorities' (Gilroy 1991: 5). Neither is the space of the diaspora the party ground for the celebration of 'the saturnalia which attends the dissolution of the essential black subject' (ibid.: 5). It might almost be said that Gilroy is looking for a third space or an excluded middle-ground between these two extremes.

So instead, the diaspora is an invocation of communal space which is simultaneously both inside and outside the West. The outcome of such positioning is a form of cultural fusion; such *syncretism* produces diaspora-specific resources of resistance, a Black sensibility which for Gilroy has the power to conflate ethics, aesthetics culture and politics by the creation of subversive new public spaces in seemingly the least propitious of circumstances. It is in such spaces that even in Victorian England touring Black musicians can subvert imperial ideologies and speak to the experiences of the white working classes in the nineteenth century as much as their heirs provide the diasporic syncretism of musical forms that inform the cultures of late twentieth century cosmopolitanism (Gilroy 1987). In short, the spatiality of the diaspora is the ground on which momentary and ever-shifting lines are drawn between inside and outside, oppressor and oppressed, the same and the other.

These lines stress interconnection as much as distinction, but they produce a space in which identities are momentarily authenticated, on which what might be called *arbitrary closure* occurs. Rejecting both essentialized and depthless representations of Black identity, Gilroy's diaspora is the spatiality which contingently mediates Black authority, in the explicit knowledge that an imagined space of diasporic identity is located within global systems that not only make such claims context-specific, but also make communication through the myriad forms of cultural syncretism inevitable. It is a stress on the connections through space and the corresponding links between places and peoples that Doreen Massey has also been particularly interested in exploring (1991; 1992; Chapter 8, this volume), a rhetoric in which the bonding of different experiences through their spatialization displaces the common implications of exclusion that the geography of *communities* can imply.

In Gilroy's work, the cultural politics of musical expression provide an exemplary case of this linking process. Reggae, soul, rap and many other genres of Black music link the Caribbean, Europe, Africa and the Americas. Such cultural forms draw on a specific, shared diasporic sense of identity but also communicate outside diasporic boundaries. So 'raga(muffin)' in the 1990s becomes a form of musical syncretism that is rooted in the Black internationalism of the diaspora but communicates beyond it. In processes of cultural fusion, it becomes a mode of the expressions of dissent in the youth clubs of the riot torn, almost all white Meadowell Estate in the north-east of England. It is also taken up by young Bengalis in the East End of London who mix it with an alternative diasporic sensibility in drawing on the Punjabi musical forms of Bhangra music.

The political significance of such expressions may be moot but their political content is manifest. What is of particular relevance here is that they rely on a cultural hybridity through which political codes of difference are crossed and transgressed through the processes of syncretism rooted in simultaneously imaginary and real spatialities. In the terms of an important article by Paul Gilroy, 'it ain't where you're from, it's where you're at' (Gilroy 1991).

The cultural politics of diaspora thus conceived also find an epistemological equivalent, seen repeatedly in the troubled ways in which social science has variously legitimated and discredited ethnically specific perspectives in the productions of *knowledge* about *race relations*. Such perspectives, sometimes described as local knowledges, have at times drawn respectability out of the postmodern critiques of the possibilities of meta-narrative certainty. Yet the possible slippage into ethnic essentialism and cultural relativism that follows defines tensions which the notion of diaspora neatly sidesteps.

So, set against the domination of the established academy, postcolonial discourse for Homi Bhabha may productively appropriate the crisis of Enlightenment reason, which is marked by the incommensurability of different articulations of identity. Typically, he has celebrated the 'conflictual articulation of meaning and place, the partial – and double-identifications of race, gender, class, generation at their point of unfamiliarity, even incommensurability' (Bhabha 1992: 60).

Bhabha goes on to suggest that his focus is 'the moment of culture caught in an aporetic, contingent position, in-between a plurality of practices that are different and yet must occupy the same space of adjudication and articulation' (ibid.: 60). This he describes as a liminal form of cultural identification. In another sense what is being expressed here is the simultaneous realization of different spatialities. Post-colonial realities routinely produce such co-presence because at root they are forever articulating Fanon's paradox that 'the Black man's soul is a white man's artefact'. This in no way should be taken as meaning that post-colonial identities are exclusively controlled by colonial legacy. But it is more mundanely the case that discourses of racialization invoke different, frequently irreconcilable identities. In British racist discourse, the 'coloured' of coloured migrants is not commensurable with the omnibus abusive epithet 'Paki' or the gendered and aged 'Black' that informs the criminalization of young Afro-Caribbean men.

It is the simultaneous presence of multiple spatialities which provides the medium through which such contradictions may be subsumed or even naturalized (Keith 1993). This suggests that epistemological problems associated with situated knowledges may in part be resolved by unpacking the spatialities that they so unproblematically evoke (see Part 2). Just as Gilroy's identity formation is tied to the imagined geography of diaspora, the epistemological incommensurability with which Bhabha is concerned is linked to the liminal spaces on which his analysis rests.

Politically, there is a reactionary vocabulary of both the identity politics of place and a spatialized politics of identity grounded in particular notions of space. It is the rhetoric of origins, of exclusion, of boundary-marking, of invasion and succession, of purity and contamination; the glossary of ethnic cleansing. But there are also more progressive formulations which become meaningless deprived of the metaphors of spatiality. Debates around territorialized and diasporic politics and political authority are just two instances where opposing the reactionary and promoting the progressive is possible only if the spatializations on which they rest are unpacked and made explicit. Such spatialities are necessarily always the source of both ethical optimism and political caution.

FROM IMAGINATION TO POLITICS TO REALITY . . . AND ALL THE WAY BACK AGAIN

The two illustrations above exemplify the identity politics of place and the spatialized politics of identity. Many of the themes touched on in these examples are central to a brilliant novel, which takes hybridization as its central dynamic, whose characters' identities are the site of struggle between purity and impurity, the sacred and the profane. It is a novel which tells of the experiences of the uprooting, disjuncture and metamorphosis that lies at the heart of the migrant condition, which itself stands as metaphor for the experiences of the whole of humanity. It is a novel about the desperate search for wholeness by its two principal protagonists, two characters who are exceptional not in their personification of hybridity and syncretism but only through the circumstances that force them to realize that they can never be made whole, or perhaps more significantly, can never be made pure: two identities which through what they lack remain in eternal search for completion. It is very much a novel about how absences are constitutive of presence, a source that Homi Bhabha has used to illustrate his notions of incommensurable identities and Stuart Hall has used to describe the emergence of the new ethnicities in late twentieth-century Britain (Bhabha 1992; Hall 1988).

The novel is *The Satanic Verses* by Salman Rushdie, a work that prompted a crisis at the very heart of liberalism, on the ground where free speech and mutual tolerance become mutually irreconcilable. In both a banal and a profound sense, what the Rushdie affair was about was the way in which a burnished, polyphonous, complex and contradictory piece of prose was transformed into an orientalist symbol in a racist society. In Part 2 of this introduction we want to suggest that both in the text and in its realization, *The Satanic Verses* can be understood in a way that usefully points towards both the theoretical problematics that lie at the heart of this collection and also to the theoretical value that a spatialized reading of the novel might accrue.

As Rushdie has said himself, 'the city as reality and the city as metaphor is at the heart of all my work' (Rushdie 1991: 404). As one of his characters puts it,

> 'The modern city' Otto Cone on his hobbyhorse had lectured his bored family at the table, 'is the *locus classicus* of impossible realities. Lives that have no business mingling with one another sit side by side upon the omnibus. One universe, on a zebra crossing, is caught for an instant, blinking like a rabbit, in the headlamps of a motor-vehicle in which an entirely alien and contradictory continuum is to be found. And as long as that's all, they pass in the night, jostling on Tube stations, raising their hats in some hotel corridor, it's not so bad. But if they meet! It's uranium and plutonium, each makes the other decompose, boom.'
>
> (Rushdie 1988: 314)

With considerable feeling, Rushdie has said 'it is hard to express how it feels to attempt to portray an objective reality and then to have become its subject' (Rushdie 1991: 404).

Running through the novel, spaces become the forces of dislocation that both make our longing *to know* that much more powerful and make our inability to do so or to judge between difference that much more difficult. At times the book satirizes Islam but it is clearly written by a secular author who seriously examines the perennial puzzle of faith and the loss of faith and who equally seriously disparages those who claim to have the answers to such mysteries, particularly if such answers are trite. There is an echo of something Rushdie himself wrote ten years ago in the novel *Shame*.

> Outsider! Trespasser! You have no right to this subject! . . . Poacher! Pirate! We reject your authority. We know you with your foreign language wrapped around you like a flag: speaking about you in your forked tongue, what can you tell but lies? I reply with more questions. Is history to be considered the property of the participants solely? In what courts are such claims staked, what boundary commissions map out their territories? Can only the dead speak?
>
> (Rushdie 1983: 28)

Both the central problem and the subsequent events surrounding *The Satanic Verses* return over and again to the vexed question of judging between different kinds of difference. But who is to play Pontius Pilate and pass judgement on Rushdie, however reluctantly?

2

INTRODUCTION PART 2

The place of politics

Michael Keith and Steve Pile

MAPPING *THE SATANIC VERSES*

Who is to play Pontius Pilate and pass judgement on Rushdie, however reluctantly? There are two routes we want to take out of our spatialized reading of *The Satanic Verses*. The first route tracks the evocations of the spatial, particularly the city as simultaneously real and imaginary, suggesting that Rushdie draws, perhaps unwittingly, on a notion that *the urban* invokes a multiplicity of spatialities simultaneously present. The second route is one that traces the notions of impurity and hybridity which are so central to the novel and suggests that this leads us into an understanding of identities defined as much by what they lack as by what they include.

The two routes draw respectively on the theoretical guidance of Henri Lefebvre and Ernesto Laclau to suggest that some of the central problems that are consistently flagged in identity politics can be resolved by mapping the spatialities on which both incomplete identities and situated knowledges rest. There is no suggestion here that Lefebvre and Laclau provide the only routes to the points we want to make, only that their work is exemplary in this context.

ROUTE 1: THE CASE FOR MULTIPLE SPATIALITIES

'The West': A description of economic development but also an imagined locus of a particular form of rationality, sometimes reified by caricature in postcolonial enquiry. Symbolically at the heart of global power, 'the West' is a linguistic condensation of the globally powerful. It is in this context that, in both orientalist imagining and in the counter memories of resistance, Islam is placed in literal and symbolic opposition to this force. Such positioning by no means exhausts a description or analysis of contemporary Islam but it is constitutive of it. Such positioning also has literal and symbolic dimensions and it was into such configurations of power and powerlessness that *The Satanic Verses* was thrown; in Rushdie's own terms, a political event that was used in the cultural politics of Indian sectarianism and then reimported, tragically signified in the homologous binaries of Islam against anti-Islam, faith against faithlessness, powerless against powerful, Good against Evil.

And whilst the symbolism with which the book was endowed had little or nothing to do with the intentions of the author, the racist configuration of the social in which such a symbolism emerged was a landscape that he knew very well. Because Rushdie, more than most authors, knows that the power of naming and describing is so context-specific.

A commonplace defence of the reactionary white is to question allegations of racism by false equivalence. 'Why', we are so often asked, 'is it alright for her to call me a Pom but racist for me to call him a nigger?' And as anybody well versed in such encounters will immediately tell you, the false equivalence arises because a particular power relation is masked by an apparently universal right to abuse. The (spatial) context is a constitutive element of the individual speech act, the specific instance of everyday racism. And Rushdie's satire was not of the powerful by the powerless, he mocked Islam from the institutionally endorsed heights of the literary establishment. Constructed in British society as a lionized Booker prize-winning novelist, he was gifted with a voice that alone drowned out the hurt protestations of the faithful, whose very faith had systematically been used historically as the medium of their degradation.

On one level this is why, whilst we would defend Rushdie's creation and his inventive genius, the novel can never be considered completely innocent; notwithstanding the grotesque asymmetry of his putative 'crime' and prescribed punishment. The iconic status of the Rushdie affair outgrew the apparent sincerity and the alleged (inevitable?) *'mauvaise foi'* of the text.

On another level, both the affair and the text illustrate some of the themes that have pervaded all of Rushdie's work and are shared by this volume: that the metaphoric and the real do not belong in separate worlds; that the symbolic and the literal are in part constitutive of one another. That meaning is never immanent, it is instead not just *marked* but also in part *constituted* by the spaces of representation in which it is articulated. These spaces of representation subvert the representation of spaces so that the ground we stand on becomes a mongrel hybrid of spatialities; at once a metaphor and a speaking position, a place of certainty and a burden of humility, sometimes all of these simultaneously, sometimes all of them incommensurably.

And so, as several chapters in this volume demonstrate, it is not possible to draw unproblematically on a spatialized vocabulary without reflecting on how this vocabulary operates. Different terms *connote* different meanings. They also mean different things at different times and different places to different people: for example, a term like 'place' triggers a chain of associations with parochialism, difference and ultimately reaction for some, and for others the term 'space' may set off a more approbatory chain tied to transcendence, universality and enlightenment.

We are arguing neither that one set of associations is valid, nor that it is possible or desirable to swap elements of the chain of signification around so the normative evaluations of space and place are reversed. It is instead suggested that we must look to the sites in which these associations are evoked (spaces of representation) in order to understand the cultural production of the representation of space.

The distinction between the representation of space ('the conceived') and the spaces of representation ('the lived') is one drawn from the work of Henri Lefebvre in *The Production of Space* (Lefebvre 1991 (French edn 1974)). Lefebvre is best known for his insightful analysis of how the twin myths of transparency and the illusion of realism represent space as a neutral and passive geometry. These myths mask the fact that space is produced and reproduced and thus represents the site and the outcome of social, political and economic struggle. It is this emphasis on the production of space that has been taken up in attempts to produce a political economy of space, most powerfully in the work of Marxist geographers such as David Harvey, Doreen Massey, Neil Smith and Ed Soja.

Lefebvre is also keen to distinguish the productions of different kinds of space, not just a distinction between the metric of space (physical space), mental space and the social space that contains appropriate place for the social relations of reproduction (gender relations) and relations of production (the division of labour) (Lefebvre 1991: 32). More interestingly, Lefebvre develops the notion of different forms of produced space, not just the deconstruction of the illusions of naturalness and transparency but instead a typology of spatialities that covers a range from sensory, sensual representational spaces through to the space of the Greek city that is assumed in classical philosophy and inscribed in ostensibly universal concepts such as citizenship.

However, in Lefebvre's work the analytical shift from 'things in space' to 'the production of space' is historical as well as logical. The insights to be gained from the focus on the production of space can be lost by a concentration on the product as artefact, ossified by either the inertia of the socio-spatial dialectic or the vagaries of rent. Conceptually squashed into the frame of reference as an artefact, space is always in danger of losing its plastic, multifaceted character. In large part this can be traced from Lefebvre's take on space as evolutionary. For him, the globe is progressively dominated by distinct forms of spatiality.

Different forms of space succeed each other through time. There is a succession from natural to absolute to abstract space, progressively erasing *nature* from our sense of spatiality. The teleological high point of capitalism is a world structured by abstract space which has 'homogeneity as its goal, its orientation, its lens' (Lefebvre 1991: 287). In turn, the contradictions of capitalist space contain within it 'the seeds of a new kind of space' (ibid.: 52). The spatial expression of contradictions can be understood as *differential space*, which hinders the workings of the system and becomes capitalism's Achilles' heel.

Situated in the site-specific politics of his own relation to the events in Paris in 1968, Lefebvre is anxious to pre-empt charges of apostasy by sticking to a rigidly teleological evolution in which absolute space is gradually erased by abstract space, which in turn gives way to differential space.

For our argument here, we want to suggest that this teleological element of Lefebvre's analysis in fact detracts from the analytical power of the work. It is not that we undervalue the significance of matching stages of capitalism with corresponding spatialities, as Harvey's work demonstrates in great detail. It is instead the case that the emphasis on succession in Lefebvre's work tends towards historicism and detracts from the conceptual richness of the notions of the spatial that Lefebvre outlines. As he himself makes clear with reference to *the urban* and the role of Paris as both crucible of conflict and container of dissent (Lefebvre 1991: 385–8, also Soja and Hooper, this volume) there are tensions and contradictions here but the overall thrust of Lefebvre's historicism is to fall prey to the myth of immanence, the notion that at each historical moment space is dominated by one (produced) set of immanent meanings.

The rigidity of this conceptual schema understates the malleability of the symbolic role of landscape. Space is produced in the image of capital but can be reappropriated in the symbolic vocabulary of liberation. Just as nationalism is challenged when the Union Jack and the Stars and Stripes are profaned. Just as 'queer' and 'nigger' are reinstated at the centre of site-specific progressive politics or the dominant representations of space satirized by the nuclear-free zone. All transform received meaning and reinvent space as the site of the frontier effects around which political mobilization occurs.

Again it is important that what we are *not* drawing from Lefebvre is a suggestion of an infinite number of forms of spatiality, all of equal significance. Such a proliferation of new spaces would itself be counter-productive. Social analysis shorn of political economy begets a politics of the Right because a specific economic order assumes the unspoken allure of the natural. In divorcing the manufacture of underdevelopment from the celebration of affluence – the icon of America – the affluent belies the reality of the Los Angeles of the oppressed. As David Harvey forcefully argues in Chapter 3, the social relations of capitalist production remain, and in comprehending their injustice it is imperative that we figuratively connect the spaces through which they are realized.

Instead we stress caution in understanding the multiplicity of spatialities simultaneously present at any one time. None of these is necessarily signified. Although de Certeau understandably chooses to celebrate the poets of the streets, the subversive consumers of urbanism, such a celebration cannot be unqualified. It has its negative potential, particularly if it emerges in exoticized representations of the city *flâneur*. Likewise, Deutsche (1990) is surely right to stress the potential will to power implicit in the view from on high that is coded into the critical distance of urban studies, but equally, wrong to imply that this is a necessary political stigmatization of such a view. As the allusion to the connotative power of the words 'place' and 'space' suggested, the processes of signification are far too complex to lend themselves to such easy good/bad classifications. It is rather that we can return to one of the protagonists in *The Satanic Verses* who tries to grasp London by knowing it but finds instead that

> the city in its corruption refused to submit to the dominion of the cartographers, changing shape at will and without warning, making it impossible for Gibreel to approach his quest in the systematic manner he would have preferred.
>
> (Rushdie 1988: 327)

ROUTE 2: 'TAKING PLACE' – THEORIZING CONTINGENCY AND RADICAL CONTEXTUALIZATION

There is a sense in which our sensibilities are spatialized. We have seen already the manner in which the postmodern turn has frequently drawn on a spatially rich vocabulary, and more specifically how Fredric Jameson turns to cognitive mapping almost as a search for security in a world of hyperreality. Similarly, for authors like Ed Soja and Umberto Eco, contemporary landscapes are, after Lefebvre, freed of material or real referents, structured by abstract rather than absolute space (Eco 1986).

So we are told that in the postmodern city, we live in spaces that, after Baudrillard, are simulacra of realities which quite possibly never existed. The space of hyperreality is the space of the true fake, a faithful imitation of something that never was. It is in such hyperreal space that the rich go lifestyle shopping, drawing on seemingly mutually exclusive benevolent significations of both town and country simultaneously, exemplified in the synthetic public spaces of spectacular regeneration and the vogue notion of the urban village. The spaces of the poor meanwhile are structured by the security-paranoid gaze which fixes bodies in carceral compounds, pre-empting insurrection.

There is both rhetorical force and intuitive appeal in such analyses of the contemporary condition. But a conflation of the spatial and the novel can mask continuities that link the notion of the true-fake with past spatialities. In one guise we have already seen how, for Walter Benjamin, identity and location were inseparable: knowing oneself was an exercise in mapping where one stands.

So, while we are not arguing against specific readings of the spatiality of the postmodern city, we are suggesting that spatialities have always produced landscapes that are loaded with ethical, epistemological and aestheticized meanings. Almost invariably these are contested. In contrast, there is an implicit acceptance of a dichotomy between myth and reality in some depictions of hyperreal space that misreads Eco's notion of the real fake. The self-consciousness of synthetic landscapes draws, frequently ironically or playfully, on lifestyle archetypes that are parodies of reality. At times the implication, as we saw in Zukin's work, is that once this was not so; that once there were real urban landscapes (relatively) lacking such artifice.

Such a stress on the contemporary echoes the flawed teleology of Lefebvre. A stronger argument is to suggest that the social, the political and the economic do not just take place in 'time' and 'space', they are in part constituted by temporality and spatiality. In this vein, Ernesto Laclau has long argued that the commonsense notion of society passing through time is illusory. Both are constitutively linked with one another, *historicity* is consequently a necessary property of all social relations.

For our part, we would argue for an equivalence here between historicity and spatiality, an assertion articulated more fully in Doreen Massey's chapter, because there is a sense in which Laclau's work is significant precisely because it rests on the rejection of the binary dualisms between the abstract and the empirical, the universal and the specific. Working within a post-Marxism that owes its lineage to the way in which hegemony works invariably as a process that escapes closure (Laclau and Mouffe 1985) and his *New Reflections on the Revolution of Our Time*, Laclau's arguments are very much with the failure of Marxist historicism to 'deliver the political goods', more specifically the resort of certain strands of Marxist thinking to explain this failure in terms of the escape clauses tied to a notion of *conjuncture*.

For Laclau, such thinking is based on a hermetic, irrefutable logic that all historical events are reducible to the abstractions of the Marxian canon except those which are not, which are the contingent effects of the vagaries of conjuncture. In both his work with Chantal Mouffe and his more recent collection of essays, Laclau works within a Marxian frame of reference to take issue with precisely this escape clause. Simplifying massively, what Ernesto Laclau argues is that all objects of scrutiny in the social sciences are incompletely constituted because of their location within a field of difference from one another. They are defined by the *negativity* that separates them and are marked with a particular *historicity*.

Historicity provides the forces of dislocation which always block the formation of complete objects but also, in blocking, affirm their very existence. Any articulation of identity or object formation is only momentarily complete, it is always in part constituted by the forces that oppose it (*the constitutive outside*), always contingent upon surviving the contradictions that it subsumes (*forces of dislocation*). In such a fragile world of identity formation and object formation, political subjects are articulated through moments of closure that create subjects as surfaces of inscription, mythical and metaphoric, invariably incomplete.

In this sense, identity emerges through difference, just as all object formation is always partial because always relational. This negativity is the source of what Laclau draws on extensively in a concept of the *constitutive outside*. This is the source of his well-known diagnosis that society can never be wholly constituted as an object of scrutiny, in the case of the social; 'the social never manages to fully constitute itself as an objective order' (Laclau 1990: 18) because of the presence of the constitutive outside.

Most usefully, here Laclau draws implicitly on Lacan to work through the concept of identity formation set within these fields of (negative) differences, being articulated through that which it is not (*negativity* in Laclau's terms) and through the historical moment of enunciation (referred to as *contingency*). Difference becomes *located* difference within a relational field: 'what one gets is a field of simply relational identities which never manage to constitute themselves fully, since relations do not form a closed system' (ibid.: 20–1). Consequently, identities and their conditions of existence are inseparable. There is no identity outside of its context: 'Identity depends on conditions of existence which are contingent, its relationship with them is absolutely necessary' (ibid.: 21).

Importantly for us, identity is always an incomplete process. At times, in order to make sense of a particular moment or a particular place (synchronic analysis), this process is stopped to reveal an identity that is akin to a freeze-frame photograph of a race-horse at full gallop. It may be a 'true' representation of a moment but, by the very act of freezing, it denies the presence of movement. The photograph represents a momentary stop in this gallop, simultaneously real and unreal, it is a moment at which *closure* occurs. Likewise, with identity, the very act of representing the ceaseless process of identity formation is based on a moment of *arbitrary closure* which in the same fashion is both true and false simultaneously. More technically, synchronic analysis is necessarily a process of sometimes justifiable misrepresentation. This is why identity is always incomplete, always subsumes a lack, perhaps is more readily understood as a process rather than an outcome.

There are also epistemological consequences of this – for Laclau, negativity reveals the contingent nature of all objectivity.

> It is not possible to threaten the existence of something without simultaneously affirming it. In this sense, it is the contingent which subverts the necessary: contingency is not the negative other side of necessity, but the element of impurity which deforms and hinders its full constitution.
> (Laclau 1990: 27)

Hence objectivity for Laclau is always partially threatened and partially constituted

> As Saint-Just said: 'What constitutes the unity of the Republic is the total destruction of what is opposed to it.' This link between the blocking and simultaneous affirmation of an identity is what we call 'contingency', which introduces an element of radical undecidability into the structure of objectivity.
> (ibid.: 21)

This notion of antagonism links directly back to the mutually constitutive nature of Islam and the West, which we have already described. Each is fixed in the imagination of the other, but this does render the two identities thus produced equivalent precisely because power relations between the two identities determine the nature of articulation.

Equally significantly, in producing a frame of analysis opposed to essentialism, Laclau talks of weakening the boundary of essence through the radical contextualization of any object (ibid.: 23). Both the objects which are discursively formed through our attempts to make sense of the world (epistemological products) and objects that are formed through our attempts to make sense of ourselves and each other (identities) are subject to this process of *radical contextualization*. In fact, it is the ability to articulate either set of objects that is the very expression of power and the moment when the political (for Laclau, that which is contested) and the social (for Laclau, those practices that are sedimented in time and uncontested) define each other by their mutual opposition: 'The constitution of a social identity is an act of power and that identity as such *is* power' (ibid.: 31, emphasis in original). An objective identity is consequently not a homogeneous point but a set of articulated elements in different discursive settings. One way Laclau phrases this is through the assertion that all social relations have contingent conditions of existence understood in terms of a *radical historicity*.

Any single set of articulated elements which defines subjects (akin to Foucauldian subjectification) is thus simultaneously an expression of power. Paradoxically, this expression of power always contains within itself a constitutive outside opposed to the particular articulation. Consequently, the inability to articulate any objectivity which is complete or without such expressions of domination prompts Laclau to use the concept of dislocation in a way that makes it central to his argument because

> every identity is dislocated insofar as it depends on an outside which both denies that identity and provides its conditions of possibility at the same time. But this in itself means that the effects of dislocation must be contradictory. If on the one hand they threaten identities, they are the foundation on which new identities are constituted.
>
> (ibid.: 39)

Laclau describes dislocation exclusively in the vocabulary of historicity, a decision which (we think mistakenly because he takes a notion of 'space as routinized time' from Heidegger) relegates the status of spatiality in the process of dislocation, something that Doreen Massey discusses in detail in Chapter 8. What we are interested in is reconceptualizing spatialities using Laclau's notion of *surfaces of inscription*, which are themselves marked by necessarily incomplete object formations. Therefore what we want to draw from Laclau – both thematically in this volume and specifically in our reconsideration of the Rushdie affair – is the stress on the *necessary* incompleteness of identity formation. As he puts it, 'the field of social identities is not one of full identities but of their ultimate failure to be constituted' (ibid.: 38).

This has major consequences for two forms of purity which we are keen to reject in this volume. The first is the notion of pure knowledges. Because all object formation is relational, all knowledge is necessarily situated: 'all objectivity necessarily presupposes the repression of that which is excluded by its establishment' (ibid.: 31). The fact that the social never manages to constitute itself fully as a social order presents particular problems for epistemology. In no way threatening notions of realist ontology, the radical step that Laclau takes is to reconnect the empirical and the abstract in such a way as to make them mutually constitutive. Consequently, any statement about what we know is itself subject to its own logic of radical contextualization. There are signs of a similar sort of project in David Harvey's call in this volume (Chapter 3) for a dialectic between the universal and the particular in the evaluation of claims made on the grounds of difference.

The second is that of pure difference. A politics of identity that stresses the irreconcilable nature of differences can, if at times only implicitly, promote a notion of a politics of location that privileges each and every location. The advantage of Laclau's theorization is that these spaces can be unpacked, they become equivalent to surfaces of inscription, which can only work through an understanding of the subject of political action. Such subjects have four properties for Laclau:

1 Any subject is a mythical subject.
2 The subject is constitutively metaphoric.
3 The subject's forms of identification function as surfaces of inscription.
4 The incomplete character of the mythical surface of inscription is the condition of possibility for the constitution of social imaginaries.

And so, in selectively combining the analysis of Laclau and Lefebvre, our understanding of identities, epistemologies and spatialities are resolved effectively through a theorization of the nature of contingency, a foregrounding of the limits to abstraction through the practice of radical contextualization.

Identity should always be a process, never an artefact. The moment the former is transformed into the latter is the point at which a contestable closure is transformed into a reified boundary. In another context, this is perhaps what Donna Haraway means when she talks of a distinction between affinities and identities where 'Feminist discourse and anti-colonial discourse are engaged in the very subtle and delicate effort to build connections and affinities and not to produce one's own or another's experience as a resource for closed narrative' (Haraway 1991: 113).

Where representation necessarily misrepresents, we find the political moment when the strategic nature of closure is revealed. Again we should not present this as something novel. Instead, such closures are moments in a politics of articulation, an echo of something Fanon reflected on while listening to the radio in the midst of anti-colonial war:

> The Arabic channel was of course jammed. But the scraps of sound had an exaggerated effect. Like rumours, they were constructively heard, and listening to them became an act of participation in revolutionary victories which might never have occurred. To quote Fanon, 'the radio receiver guarantied this true lie'.
>
> (Feuchtwang 1990)

Which brings us full circle back to Rushdie's protagonists, Saladin Chamcha and Gibreel Farishta – colonial exiles exploring themselves as they explore London, and discovering the nature of the true-lie. Because 'most migrants learn and can become disguises. Our own false descriptions to counter the falsehoods invented about us, concealing for reasons of security our secret selves' (Rushdie 1988: 49). Though more poetic, this is surely close to Laclau's suggestion that it is the necessarily incomplete character of the mythical surface of inscription that is the condition of possibility for the constitution of social imaginaries.

DESTINATION: MULTIPLE SPATIALITIES, RADICAL CONTEXTUALIZATION, INCOMPLETE IDENTITIES, AMBIVALENT EPISTEMOLOGIES

Rethinking the Rushdie affair helps us to understand the manner in which 'the place of politics' is always itself an invocation of spatiality. It also provides a way of rethinking identity politics, not as some sort of surface froth that floats around on top of more important social processes, but as something that strikes deep into our ability to transform the social world into concrete knowledges.

To bring together the routes we have taken out of *The Satanic Verses*, we want to suggest that by combining notions of multiple spatialities simultaneously present with the practice of radical contextualization, we can both understand the significance of spatiality in the incomplete process of identity formation and reject the relativism that often bedevils identity politics (see Bondi, Chapter 5).

For us, a politics of identity is quite specifically not about an epistemology of relativism. We stand with Donna Haraway in suggesting that

> The alternative to relativism is partial, locatable, critical knowledges sustaining the possibility of webs of connections called solidarity in politics and shared conversations in epistemology. Relativism is a way of being nowhere while claiming to be everywhere equally.
>
> (Haraway 1991: 191)

The problem of spatiality is that when a gaze from nowhere becomes a gaze from somewhere, it is possible to forget at times quite how problematic that somewhere actually might be. As a modesty suit constraining epistemological boasting, this is a fine act of public humility, but unless the sorts of spatiality that are being evoked are examined more closely, we have to ask whether it is more than rhetorical gesture. A proliferation of sites of difference begets a Babel-like world where truth claims, ethical claims, and assertions of desire all offer no external criteria of refutation. There is an almost sensual challenge to 'the order of things', a danger that in articulating complexity we celebrate incoherence.

Instead, we are moving towards a sense in which vocabularies of (singular) space are unpacked in terms of the (plural) spatialities that they connote, and consequently our notion of a singular geography is displaced by (plural) geographies, avoiding relativism through the theoretical refusal to submit to the separation of the abstract and the empirical in the process of theorization.

Geographical analysis consequently rests on exactly the sort of key themes of liminality, enunciation and articulation that we have explored in this introduction.

> The epistemological distance between subject and object, inside and outside, that is part of the cultural binarism that emerges from relativism is now replaced by a social process of enunciation. If the former focuses on function and intention, the latter focuses on signification and institutionalization. If the epistemological tends towards a 'representation' of its referent, prior to performativity, the enunciative attempts repeatedly to 'reinscribe' and relocate that claim to cultural and anthropological priority (High/Low; Ours/Theirs) in the act of revising and hybridizing the settled, sententious hierarchies, the locale and the locutions of the cultural.
>
> (Bhabha 1992: 57)

Again, the sort of geographical project we are arguing for can be illustrated by returning to *The Satanic Verses*. The novel's representation of space is in part inseparable from the representational space through which the text becomes 'The Rushdie Affair'. Significantly, it is a book which has also been drawn on to inform writings about the 'new ethnicities' that are emerging in cosmopolitan societies where the link between ethnicity and nation is cut and culture retains an autonomy of sorts in the construction of hybrid new ethnicities among migrant groups.

Rushdie has argued that the novel is a 'privileged arena' (Rushdie 1991: 429). On these terms, *The Satanic Verses* is unimpeachable. But this arena is itself a product; to confine consideration within its privileged boundaries is to tell only half the story. It is worth returning to Lefebvre's distinction between the representation of space and spaces of representation.

In semiotic terms, there is a paradigmatic equivalence between the space of the academy, in which the vogue for identity politics is articulated, and the representational space of the novel. These spaces of representation are inscribed with ethical and epistemological, as well as aesthetic, traces and conventions. It would be obscenely absurd to blame Rushdie for the cynical manipulation of the symbolic power of the novel by politicians of this and other countries. But there is another absurdity, equally extreme, equally common. And that is to pretend that the scriptor/author is ignorant of the social context in which the work is produced, the potential for a novel that draws, *however sympathetically*, on the life of the prophet to become an orientalist icon in a racist world. This is the representational space in which the text 'takes place'. *The Satanic Verses*, regardless of the author's inclinations, is marked by the stigmata of the literary world that defines it as a classic.

Likewise, in the recent work of bell hooks, Stuart Hall, Paul Gilroy, Homi Bhabha and others, descriptions of the hybridity of contemporary cultural politics invoke only a sympathetic *celebration* of the syncretism that is happening 'out there' in the real world. Yet in defining new ethnicities, there is clearly potential for the texts to signify a rejection of 'old' ethnicities; not only the exoticizing classificatory gaze of the ivory tower's old ethnicities but also the meaning of these 'old' identities in other worlds. In the latter context, there is clearly a potential (mis)reading of the new identity politics indicted for devaluing the cultural heritage that resources resistance, just as there is a potential (mis)reading which suggests that the work on new ethnicities offends because it transgresses old unitary mobilizations of political blackness/class identity (Sivanandan 1990).

Political, ethical and aesthetic significations are found in the representational spaces that are outside the hermetic world of the literary or academic text. As Patricia Williams points out, openness and syncretism are not universally 'good', they themselves are structured by the context in which they take place: in a white hegemonic world, there is also 'openness as a profane relation. Not communion, but exposure, vulnerability, the collapse of boundary in the most assaultive way' (Williams 1991: 199).

In one sense what we have here is a return to the myth of spatial immanence and a fallacy of spatial relativism referred to in Chapter 1. In an abstract literary world, such concerns appear self-evident. Texts neither have immanent meanings, nor are their meanings entirely the creation of the audience. Yet it is the spatial articulation of these problems that is so confusing.

Benjamin again is useful here precisely because in his representation of space, his Berlin and Paris were not only simultaneously real and metaphorical worlds, they were also acts of representation that were consciously, cognitively and politically marked rather than the evocations of a purely aesthetic spatiality. However, as feminist critics have long pointed out, the *flâneur* was a mode of sensibility, a viewpoint on flux, a position sometimes disingenuously acknowledged (Pollock 1988). Such a viewpoint may or may not have been open to women in the city (Wilson 1991). But more programmatically, it is only if both the spaces of representation and the spatialized vocabulary (representations of space) of contemporary social theory are rendered explicit that we can move towards the project that Laclau describes as 'radical contextualization'.

In relational terms, we want to move away from a position of privileging positionality and towards one of acknowledging spatiality. Such a move takes us towards an understanding of identities as always contingent and incomplete processes rather than determined outcomes, and of epistemologies as situated and ambivalent rather than abstract and universal. It is an acknowledgement of difference that gives no concessions to relativism. The ethical and political agenda can remain structured by social justice, but it is a justice that radically contextualizes the various forms of oppression to find the ground on which progressive action can be taken. This radical geography does not – like Pontius Pilate – wash its hands of judgement. It is precisely this ground that is just beginning to be explored by both a form of new jurisprudence that asks how to arbitrate between difference (exemplified by the work of Patricia Williams (1991)) and moral philosophies that make explicit the representational spaces on which they are based (exemplified in the work of Iris Marion Young (1990)).

ARENAS OF DEBATE (IN THIS COLLECTION)

It is the effort to explore and confront these all too frequently implicit forms of spatiality that is embarked upon in all the chapters in this volume. In this final section of our introduction, we would like to pick out three themes which we believe the contributors to this book develop. As you will find, the authors cut across similar kinds of issues, though in rather different ways; each issue relates to aspects of *Place and the Politics of Identity*. In many ways, it is arbitrary to isolate these aspects as if they were in some way theoretically or politically unrelated. We would therefore like to describe them as 'surfaces of articulation' because we want to imply a dynamic sense of both difference and incommensurability and also of mutual constitution and interconnection. These surfaces are both separable and inseparable; we have labelled them 'locations of struggle', 'communities of resistance', and 'political spaces'.

Locations of struggle

Perhaps fundamentally, there is a question about the constitution of the person, and how people enter as individuals into politics. The assumptions that are made about how people are constituted have profound effects not only on the kinds of radical politics people can be expected to make but also on the kind of effects that can ultimately be hoped for through political action. Harvey asks this question openly and confronts directly the problems relating to a politics of identity and difference. He is particularly worried by the lack of a political response to a major industrial accident. Here, indeed, is a location without struggle. The question, as was Lenin's, is: 'What is to be done?'

The problem is that there are different and opposing answers, crudely, based either on a return to a singular identity around which radical politics is mobilized or on deploying multiple identities strategically based on a reading of the most radical tactics in any given situation. The answers provided depend on the assessment of the current state of politics. For example, Soja and Hooper (Chapter 10) argue that there has been a fragmentation of modernist identity politics, and that this is 'an endemic problem'.

There appears to be broad agreement throughout the book on the undesirability of essentialist notions of the individual; this does not settle the issue, unfortunately. Questions then develop about how the individual is understood to be placed – located – in society (especially Bondi, Golding, Harvey, Revill, Smith and Katz). Are there social relations which people share? If people do share oppressions, are some more fundamental than others? Is it possible to ignore differences in order to form alliances against the powers that be? Which differences are to be articulated, and which are to be left for a later struggle? Around what points – moments, surfaces, events – are people to be mobilized?

This collection suggests that these questions are now being answered through appeals to the spatial – whether real spaces, imaginary spaces, or symbolic spaces. Yet notions of the spatial, and their political consequences, are not fully comprehended or fleshed out (Massey, Smith and Katz). Nevertheless, it is accepted that the individual has to be located within the struggle somehow (Bondi, Golding, Harvey). There are problems here though. Where identity is assumed to be fixed and singular, then this provides a firm base on which to mobilize politics; the down-side is that this can all too easily exclude potential allies and may be unable to adapt to changing circumstances. On the other hand, where identity is assumed to be multiple, then this facilitates a kind of guerrilla warfare against the powerful, and it authorizes all kinds of alliances and tactics; unfortunately, this may be unable to distinguish between important and irrelevant struggles, and it may create counter-productive alliances between groups who should not be 'bedfellows'.

The deployment of spatial metaphors are used to resolve these questions, in part at least by refusing a simple 'us' versus 'them' binary ordering of resistance (Golding, Soja and Hooper). The consequence of this is an examination throughout the book of locations of struggle – deploying a sense that these are simultaneously real, imaginary and symbolic – in order to create and sustain communities of resistance (Radcliffe).

Communities of resistance

This argument about the location of the individual in struggle and the commonsense deployment of spatial metaphors has led the authors to challenge what may be called hegemonic constructions of space. It is argued that there is a danger implicit in drawing on received notions of space: any mobilization grounded in reconstituting the spatial may have unintended or even reactionary elements (Radcliffe). Worse, a spatialized politics can be used for the vilest of ends and this can be seen all over the world, but perhaps especially in Serbia's policy of ethnic cleansing – so reminiscent of Nazi Germany's extermination of recidivists, criminals, Jews and other so-called 'sub-humans', homosexuals, political opponents, and others whom the Nazis labelled anti-social (see Theweleit 1989, Chapter 2).

It is argued here that thought needs to be given to the political deployment of (real, imagined, symbolic) space, and that the purpose of such questioning is to enable the formation and maintenance of progressive political alliances. This may mean the consolidation of old communities of resistance or perhaps the creation of alternative political possibilities (Harvey, Smith and Katz). The effort is to re-vision radical subjectivity and communities of resistance through 'simultaneously real and imagined geographies' (Soja and Hooper). In order to empower alliances between marginalized people, a different sense of space needs to be invoked – no longer static and passive, no longer devoid of politics (Massey).

For example, Massey's analysis shows how the commonsense dichotomy between space and time actually permits one side of the dualism to be valued and the other side to be seen as lacking value. One aspect of the exercise of hegemonic power is that it facilitates and relies on the transcoding of value between dualisms. In this way, time is valued because it is aligned with mind, reason, masculinity and – importantly – progressive politics. Whereas space is seen as lacking because it is associated with the body, emotion, femininity and – importantly – deadness (also Smith and Katz). Such a prescription denies the importance of space in the construction of radical politics; it denies both its radical and its reactionary tendencies, it denies progressive alliances constructed through the spatial.

The sense here is that space is more than the outcome of social relations and more than one of the dimensions through which the social is constructed. It is an active, constitutive, irreducible, necessary component in the social's composition. The fabric of space now becomes more than a flat two-dimensional surface. Now, it may be that space-time is four or more dimensional (Massey, Golding), it may be that there is more to space than merely being in the centre or confined to the margin (Soja and Hooper). Indeed, space can be seen to be full of gaps, contradictions, folds and tears. Through these, marginalized communities may be able to inscribe themselves into new geographies (Hesse). The marginalized may be able to make new investments, for example, in the space of Britishness (Hesse) or in the places of Argentineness (Radcliffe). From this perspective, radical politics may be seen as the effort to change the stories told about contested spaces (Hesse, Revill).

As described above, Lefebvre argues that there is a dialectic in the lived world between spaces of representation and representation of spaces. Both Hesse and Revill show that this leads to a fluidity and compositionality which escapes from the codification of the past and imagines a different future. Indeed, story-telling may be seen as a particularly effective way of crossing boundaries and building alliances – though this also contains dangers of the systematic misrepresentation of the powerless by the powerful (Smith and Katz). From this perspective, it is necessary to declare problematic any stable sense of place, politics and identity and the communities of resistance that rely on them. This need not mean that 'anything goes' as far as place, politics and identity is concerned, nor that communities of resistance are impossible. It is a recognition that stability is a struggle to achieve, and different groups have different resources which give them different capacities to articulate their position, their politics, their identities and to mobilize communities of resistance (Revill). It is a recognition of the perpetual need to create, conserve and re-create political spaces.

Political spaces

Space can now be recognized as an active constitutive component of hegemonic power: an element in the fragmentation, dislocation and weakening of class power (Harvey), both the medium and message of domination and subordination (Massey). It tells you where you are and it puts you there. The authors here agree that this is not where we want to be. The problem is this: where do we want to be, and how do we want to get there? What kind of political spaces are there to be occupied? And who is this 'we' anyway?

The authors, working through these problems, arrive at different answers. More properly, they are beginning to ask different questions. Should there be one basic organizing principle of struggle, which mirrors the fundamental structure of oppression in society (Bondi, Harvey)? Should there be multiple points of resistance, where one or more is prioritized on the grounds of their importance – or effectivity – in that situation (Massey, Smith and Katz)? Should the powerful join the oppressed in the margins (Soja and Hooper)? Should the marginal present themselves to the centre (Hesse, Radcliffe, Revill)? How should those who exist in an impossible space – because they are excluded from both the centre and the margin – act (Golding)?

It is clear that the contributors to this volume believe that some sort of position has to be adopted. The answer to the question 'Should there be multiple points of resistance, where none is prioritized?', cannot be 'Yes'. Many of the authors agree with Jane Gallop when she argues that 'identity must be continually assumed and immediately called into question' (Gallop 1982: xii). And, if the surface of politics is not flat (i.e. if all positions are not the same), then what is to be the ground on which questions of politics and identity are to be decided? And what kind of ground is this to be? An example is provided by Harvey. He argues that the left needs to recover the language of social justice – which contains a universalizing notion of human rights – while at the same time recognizing that notions of social justice are embedded in material and hegemonic circumstances. Universal notions must be situated: 'there can be no universal conception of justice to which we can appeal as a normative concept to evaluate some event'. Concepts of social justice can only provide the terrain on which progressive politics can be grounded if it is known where that terrain is and if it enables the identification of potential alliances on the basis of similarity rather than sameness.

Similarly, Hesse argues for the construction of new spaces which are not reducible to inside and outside. This also describes a movement not only from fixed to contingent surfaces of the articulation of politics, but also away from closure and totalities, partly because this enclosure is impossible as the history of diaspora demonstrates. While Soja and Hooper describe a politics that ranges from little tactics (in the lived world) to great strategies (geopolitics); in an attempt to create what hooks calls spaces of radical openness (see Part 1), Golding tries to open up 'impossible spatiality' in the name of a radical geography, which is based in having to recognize that there is an excluded middle between the same and the other, as her discussion of queer politics in Chapter 11 demonstrates.

Radcliffe shows that the Madres de Plaza de Mayo (the mothers of Plaza de Mayo) had created an alternative geography through transgression; they had come from the margins to the centre and in doing so had created a new space of resistance to the military authorities. Understanding that this was a specific challenge to a specific authority means that the mothers should not be castigated for not being radical enough, because they did not challenge hegemonic constructions of motherhood and domesticity. This example we believe is not exceptional; such an analysis may be extended to a British context: for example, the Greenham Common women who occupied public land next to a US airbase to protest against nuclear weapons, or the druids who wish to reclaim Stonehenge for their rituals.

We believe that this book demonstrates that all spatialities are political because they are the (covert) medium and (disguised) expression of asymmetrical relations of power. None of the authors simply celebrates or condemns transgression – the movement from one (political) place to another (political) place. Instead, there is commitment to a continual questioning of location, movement and direction. And so, this volume challenges – each author from his or her own perspective – hegemonic constructions of place, of politics and of identity.

REFERENCES

Ambrose, P. (1986) *Whatever Happened to Planning?*, London: Methuen.

Benjamin, W. (1979) *One Way Street*, London: New Left Books.

Berman, M. (1988) *All that is Solid Melts into Air: the Experience of Modernity* (2nd edn), London: Verso.

Bhabha, H. (1992) 'Postcolonial authority and postmodern guilt', in L. Grossberg, C. Nelson and P. Treichler (eds) *Cultural Studies*, London: Routledge, 55–66.

Brownill, S. (1990) *Developing London's Docklands: Another Great Planning Disaster?*, London: Paul Chapman.
Coupland, A. (1992) 'Docklands: dream or disaster?' in A. Thornley (ed) *The Crisis of London*, London: Routledge, 149–62.
Daniels, S. (1989) 'Marxism, culture, and the duplicity of landscape', in R. Peet and N. Thrift (eds) *New Models in Geography: Volume 2*, London: Unwin Hyman, 196–220.
de Certeau, M. (1984) *The Practice of Everyday Life*, Berkeley: University of California Press.
Deutsche, R. (1990) 'Men in space', *Artforum*: 21–3.
Eco, U. (1986) *Travels in Hyperreality: Essays*, London: Picador.
Evans, D. and Gohl, S. (1986) *Photomontage: a Political Weapon*, London: Gordon Fraser Gallery Limited.
Feuchtwang, S. (1985) 'Fanon's politics of culture: the colonial situation and its extension', *Economy and Society* 14(4): 450–73.
Gallop, J. (1982) *Feminism and Psychoanalysis: the Daughter's Seduction*, London: Macmillan.
Gilroy, P. (1987) *There Ain't No Black in the Union Jack*, London: Hutchinson.
——— (1991) '"It ain't where you're from it's where you're at". The dialectics of diasporic identification', *Third Text* 13 (Winter): 3–16.
Hall, C. (1992) *White, Male and Middle Class: Explorations in Feminism and History*, Cambridge: Polity Press.
Hall, S. (1988) 'New ethnicities' in *Black Film, British Cinema*, London: Institute of Contemporary Arts Documents, 7: 27–31.
Haraway, D. (1991) *Simians, Cyborgs and Women: The Reinvention of Nature*, London: Free Association Books.
Hardy, D. (1983) *Making Sense of London's Docklands: Processes of Change*, Geography and Planning Paper No. 9, Middlesex Polytechnic.
hooks, b. (1991) *Yearning: Race, Gender, and Cultural Politics*, London: Turnaround.
Jacobs, J. (1961) *The Death and Life of Great American Cities*, Harmondsworth: Peregrine.
Jameson, F. (1984) 'Postmodernism, or the cultural logic of late capitalism', *New Left Review* 146: 53–92.
——— (1991) *Postmodernism, or the Cultural Logic of Late Capitalism*, London: Verso.
Keith, M. (1988) 'Racial conflict and the "No-Go Areas" of London', in J. Eyles and D. M. Smith (eds) *Qualitative Methods in Human Geography*, Cambridge: Polity Press, 39–48.
——— (1993) 'From punishment to discipline? Racism, racialization and the policing of social control', in M. Cross and M. Keith (eds) *Racism, The City and the State*, London: Routledge, 193–209.
Laclau, E. (1990) *New Reflections on the Revolutions of our Time*, London: Verso.
Laclau, E. and Mouffe, C. (1985) *Hegemony and Socialist Strategy: Towards a Radical Democratic Politics*, London: Verso.
Lash, S. and Friedman, J. (1992) 'Introduction', in S. Lash and J. Friedman (eds) *Modernity and Identity*, Oxford: Blackwell, 1–30.
Lefebvre, H. (1991) *The Production of Space* (1st edn 1974), Oxford: Basil Blackwell.
Massey, D. (1991) 'A global sense of place', *Marxism Today*, June: 24–9.
——— (1992) 'Space, place and gender', *LSE Magazine*, Spring: 32–4.
Nairne, S., in collaboration with G. Dunlop and J. Wyver (1990) *State of the Art: Ideas and Images in the 1980s*, London: Chatto & Windus in collaboration with Channel Four Television Company Limited.
Pile, S. (1990) 'Depth hermeneutics and critical human geography', *Environment and Planning D: Society and Space* 8(2): 211–32.

Pile, S. and Rose, G. (1992) 'All or nothing? Politics and critique in modernism and postmodernism', *Environment and Planning D: Society and Space* 10(2): 123–36.
Pollock, G. (1988) *Vision and Difference: Feminism, Femininity and the Histories of Art*, London: Routledge.
Punter, J. (1992) 'Classic carbuncles and mean streets: contemporary urban design and architecture in central London', in A. Thornley (ed.) *The Crisis of London*, London: Routledge.
Rushdie, S. (1983) *Shame*, Harmondsworth: Penguin.
—— (1988) *The Satanic Verses*, Harmondsworth: Penguin.
—— (1991) *Imaginary Homelands: Essays and Criticism 1981–1991*, London: Granta Books.
Sivanandan, A. (1990) 'All that melts into air is solid; the hokum of New Times', *Race and Class* 31(3): 1–30.
Soja, E. (1989) *Postmodern Geographies*, London: Verso.
Sontag, S. (1979) 'Introduction', in W. Benjamin *One-Way Street*, London: New Left Books, 7–28.
Theweleit, K. (1989) *Male Fantasies II. Male bodies: Psychoanalyzing the White Terror*, Cambridge: Polity Press.
Wallis, B. (ed.) (1991) *If You Lived Here: The City in Art, Theory, and Social Activism, a Project by Martha Rosler*, Seattle: Bay Press.
Widgery, D. (1991) *Some Lives: a GP's East End*, London: Sinclair-Stevenson.
Williams, P. (1991) *The Alchemy of Race and Rights: Diary of a Law Professor*, Cambridge, Mass.: Harvard University Press.
Wilson, E. (1992) 'The invisible flâneur', *New Left Review* 191: 90–110.
Young, I. M. (1990) 'The ideal of community and the politics of difference', in L. J. Nicholson (ed.) *Feminism/Postmodernism*, London and New York: Routledge, 300–23.
Zukin, S. (1991) *Landscapes of Power: from Detroit to Disney World*, Berkeley: University of California Press.
—— (1992) 'Postmodern urban landscapes: mapping culture and power', in S. Lash and J. Friedman (eds) *Modernity and Identity*, Oxford: Blackwell, 221–47.

3

CLASS RELATIONS, SOCIAL JUSTICE AND THE POLITICS OF DIFFERENCE

David Harvey

It is hard to discuss the politics of identity, multiculturalism, 'otherness' and 'difference' in abstraction from material circumstances and from political project. I shall, therefore, situate my discussion in the context of a particular problematic – that of the search for a 'socially just' social order – within the particular material circumstances prevailing in the United States today.

HAMLET, NORTH CAROLINA

In the small town of Hamlet, North Carolina (population approximately 6,000), there is a chicken processing plant run by Imperial Foods. Chicken production is big business in these times for it can now be mass-produced under low-cost conditions of industrialized management. For many of America's poor (hit by declining incomes these last two decades), it has consequently become a major source of protein; consumption doubled in the 1980s to equal that of beef. The conditions prevailing within the broiler chicken industry, stretched in a vast arc running from Maryland's Eastern Shore through the Carolinas and across the deep south into the Texas Panhandle (the zone known as 'The Broiler Belt' because agricultural incomes are dominated by the industry) are, however, less than salubrious (salmonella contamination is an endemic danger and descriptions of production conditions are liable to stir the ire of those only mildly sensitive to animal rights). Ancillary to broiler-chicken production is a chicken processing industry employing 150,000 workers in 250 or so plants, mostly located in very small towns or rural settings throughout the 'Broiler Belt'.

On Tuesday, 3 September 1991, the day after the United States celebrated its 'labor day', the Imperial Foods plant in Hamlet caught fire. Many of the exit doors were locked. Twenty-five of the 200 workers employed in the plant died and a further 56 were seriously injured.

It was a cataclysmic industrial accident, at least by the standards of any advanced industrial country, but it also revealed (as Struck (1991), one of the few journalists to investigate, discovered) some very harsh truths about the 'latest industry of toil to reign in the [American] South'. Those employed in the plant start off at minimum wage ($4.25 an hour) and later progress to $5.60 an hour which translates into take-home pay of less than $200 per week, which is below the poverty line for a single-headed household with children. But there is little or no alternative employment in Hamlet; and for this particular town, the plant is a vital economic asset precisely because 'for a lot of people, any kind of job is better than no job at all'. Those living in relatively geographically isolated rural towns of this sort are, consequently, easy prey for an industry seeking a cheap, unorganized and easily disciplined labour force. Struck continues his account thus:

> The workers at the Imperial Foods plant describe demeaning conditions with few benefits and no job security. They were routinely cursed by bosses, the employees say. They were allowed only one toilet break from the processing line. A single day off required a doctor's permission. Any infraction was noted as an 'occurrence' and five occurrences would get a worker fired. 'The supervisors treated you like nothing, and all they want you to do is get their chicken out' said Brenda MacDougald, 36, who had been at the plant two years. 'They treated people like dogs,' said a bitter Alfonso Anderson. Peggy, his wife of 27 years, died in the fire. She had worked there for 11 years, despite her complaints. 'Around here, you have to take some stuff and swallow it to keep a job,' he said, fighting back tears.
>
> (Struck 1991)

North Carolina as a state has long had the habit of openly touting low wages, a friendly business climate, and 'right-to-work' legislation which keeps the unions at bay as the bait to pull in more and more manufacturing employment of exactly this sort. The poultry industry as a whole is estimated to add more than $1.5 billion annually to North Carolina's economy. In this case, however, the 'friendly business climate' translates into not enforcing laws on occupational health and safety. North Carolina 'has only 14 health inspectors and 27 safety inspectors (ranking) lowest in the nation in proportion to the number of inspectors (114) recommended under federal guidelines'. Federal personnel are supposed, under Congressional mandate, to make up the difference, but none have visited the plants in North Carolina in recent years. The Hamlet plant had not, therefore, been inspected in its 11 years in operation. 'There were no fire extinguishers, no sprinkler system, no safety exit doors.' Other plants in the state have rarely been inspected let alone cited for violations, though fires have been common and the occupational injury rate in the industry is nearly three times the national average.

There are a number of compelling reflections which this incident provokes. First of all, this is a *modern* (i.e. recently established) industry, whose employment conditions could easily be inserted as a description into Karl Marx's chapter on 'The Working Day' in *Capital* (first published in 1867) without anyone noticing any fundamental difference. It surely bodes ill (in some sense or other) for the 'free market triumphalism' to which we are currently exposed when looking towards the East that such a miserable equation can so easily be made in the West between nineteenth-century levels of exploitation in Britain and employment conditions in a recently established industry in the most powerful advanced industrial capitalist country in the world. The most obvious comparison in the United States is with the Triangle Shirtwaist Company fire of 1911, in which 146 employees died and which led over 100,000 people to march down Broadway in protest and which became the *cause célèbre* for the labour movement to fight for better workplace protection. Yet, as Davidson (1991) notes, 'despite a dizzying matrix of laws, regulations and codes enacted to protect workers, most of the Imperial workers died as the women in New York had: pounding desperately on locked or blocked fire doors.'

The second reflection is that we should pay close attention to the industrial structures developing in rural and small-town settings in the United States, for it is here that the decline of agricultural employment (to say nothing of the rash of farming bankruptcies) over the past decade or so has left behind a relatively isolated industrial reserve army (again, of the sort that Marx described so well in *Capital* – see Chapter 25, section 5, for example) which is far more vulnerable to exploitation than its urban counterpart. American industry has long used spatial dispersal and the geographical isolation of employees as one of its prime mechanisms of labour control (in industries like chicken processing and meat packing the equation is obvious, but this principle is also deployed in electronics and other supposedly ultra-modern industries). But recent transformations in industrial organization, flexible locational choices and deregulation have here been turned into a totally unsubtle form of coercive exploitation which is pre- rather than post-Fordist in its organizational form.

This leads to a third reflection concerning the dismantling, through deindustrialization and industrial reorganization over the last two decades, of many of the forces and institutions of 'traditional' (e.g., blue-collar and unionized) working-class forms of power. The dispersal and creation of many new jobs in rural settings has facilitated capitalist control over labour by searching out non-unionized and pliable workforces. The manufacturing sectors of central cities, which have always been more vulnerable to expressions of organized discontent or political regulation, have been reduced to zones of either high unemployment (cities like Chicago, New York, Los Angeles and Baltimore have seen their traditional blue-collar manufacturing employment cut in half in the last 20 years) or unorganized sweatshop-style industries. The non-financial zones of inner cities, which have quite rightly been the focus of so much attention in the past, have increasingly become, therefore, centres of *un*employment and *oppression* (of the sort which led to the recent explosion in Los Angeles) rather than centres of labour *exploitation* and working-class political organization of the classic sort.

But the immediate matter I wish to concentrate attention upon is the general lack of political response to this cataclysmic event. For while the Triangle Shirtwaist Company fire provoked a massive protest demonstration at the beginning of the twentieth century in New York City, the fire in Hamlet, North Carolina, at the end of the twentieth century, received hardly any media or political attention, even though some labour groups and political organizations (such as Jackson's Rainbow coalition) did try to focus attention upon it as a matter of ethical and moral urgency. The interesting contrast, in September 1991, was with the Clarence Thomas Supreme Court nomination hearings which became a major focus for a great deal of political agitation and action as well as of media debate. These hearings, it should be noted, focused on serious questions over race and gender relations in a *professional* rather than *working-class* context. It is also useful to contrast events in Hamlet, North Carolina, with those in Los Angeles, in which *oppression* as expressed in the beating of Rodney King on a highway and the failure to convict the police officers involved, sparked a virtual urban uprising of the underprivileged, while the deaths of 26 people through *exploitation* in a rural factory setting provoked almost no reaction at all.

Those contrasts become even more significant when it is realized that of the 25 people who died in the Hamlet fire, 18 were women and 12 were African-American. This is not, apparently, an uncommon profile of employment structure throughout the 'Broiler Belt', though Hispanics would typically substitute for African-Americans in the Texas Panhandle sector in particular. The commonality that cuts across race and gender lines in this instance is quite obviously that of class, and it is hard not to see the immediate implication that a simple, traditional form of class politics could have protected the interests of women and minorities as well as those of white males. And this in turn raises important questions of exactly what kind of politics, what definition of social justice and of ethical and moral responsibility, is adequate to the protection of such exploited populations, irrespective of their race and gender. The thesis I shall explore here is that it was raw class politics of an exploitative sort which created a situation in which an accident (a fire) could have the effects it did. For what happened in Hamlet, North Carolina, Struck surmised, was 'an accident waiting to happen'.

Consider, first, the general history of workplace safety and of regulatory practices and enforcements in the United States. Labour struggles around events such as the Triangle Shirtwaist Company fire put occupational safety and health very much upon the political agenda during the 1920s, and it was a fundamental feature of Roosevelt's New Deal coalition (which included the labour unions) to try to satisfy some minimum requirements on this score without alienating business interests. The National Labor Relations Board acquired powers to regulate class conflict in the workplace (including conflicts over safety) as well as to specify the legal conditions under which unions (which would often take on health and safety issues directly) could be set up.

But it was not until 1970 that a Democratically controlled Congress consolidated the bits and pieces of legislation that had accumulated from New Deal days onwards into the organization of the Occupational Safety and Health Administration (OSHA) with real powers to regulate business practices in the workplace. This legislation was, it should be noted, part of a package of reforms which set up the Environmental Protection Agency, the Consumer Product Safety Commission, the National Traffic Safety Commission, and the Mine and Safety Health Administration, all of which signalled a much greater preparedness of a Democratically controlled Congress in the early 1970s to enact legislation (in spite of a Republican President) that would extend state powers to intervene in the economy.

I think it important to recognize the conditions which led the Democratic party (a political party which, from the New Deal onwards, sought to absorb but never to represent, let alone become an active instrument of, working-class interests) to enact legislation of such an interventionist character. The legislation was not, in fact, an outcome of the class and sectional alliance politics which had created the New Deal, but came at the tail-end of a decade in which politics had shifted from universal programmes (like social security) to specially targeted programmes to help regenerate the inner cities (e.g. Model Cities and federally funded housing programmes), take care of the elderly or the particularly impoverished (e.g. Medicare and Medicaid), and target particular disadvantaged groups in the population (e.g. Headstart and Affirmative action). This shift from universalism to targeting of particular groups inevitably created tensions between groups and helped fragment rather than consolidate any broader sense of a progressive class alliance. Each piece of legislation that emerged in the early 1970s appealed to a different group (unions, environmentalists, consumer advocacy groups, and the like). Nevertheless, the net effect was to create a fairly universal threat of intervention in the economy from many special interest groups and in certain instances – OSHA in particular – in the realm of production.

The latter is, of course, very dangerous territory upon which to venture. For while it is accepted, even by the most recalcitrant capitalist interests, that the state always has a fundamental role in ensuring the proper functioning of the market and respect for private property rights, interventions in that 'hidden abode' of production in which the secret of profit-making resides, is always deeply resisted, as Marx (1967) long ago pointed out, by capitalist class interests. This treading on the hallowed ground of the prerogatives of business provoked an immediate political response. Edsall early on spotted its directions:

> During the 1970s, business refined its ability to act as a class, submerging competitive instincts in favor of joint, cooperative action in the legislative arena. Rather than individual companies seeking only special favors . . . the dominant theme in the political strategy of business

> became a shared interest in the defeat of bills such as consumer protection and labor law reform, and in the enactment of favorable tax, regulatory and antitrust legislation.
>
> (Edsall 1984: 128)

In acting as a class, business increasingly used its financial power and influence (particularly through political action committees) during the 1970s and 1980s effectively to capture the Republican party as its class instrument and forge a coalition against all forms of government intervention (save those advantageous to itself) as well as against the welfare state (as represented by government spending and taxation). This culminated in the Reagan administration's policy initiatives which centred on an:

> across-the-board drive to reduce the scope and content of the federal regulation of industry, the environment, the workplace, health care, and the relationship between buyer and seller. The Reagan administration's drive toward deregulation was accomplished through sharp budget cuts reducing enforcement capabilities; through the appointment of anti-regulatory, industry-oriented agency personnel; and, finally, through the empowering of the Office of Management and Budget with unprecedented authority to delay major regulations, to force major revisions in regulatory proposals, and through prolonged cost-benefit analyses, to effectively kill a wide range of regulatory initiatives.
>
> (Edsall 1984: 217)

This willingness of the Republican party to become the representative of 'its dominant class constituency' during this period contrasted with the 'ideologically ambivalent' attitude of the Democrats, which grew out of 'the fact that its ties to various groups in society are diffuse, and none of these groups – women, blacks, labour, the elderly, Hispanics, urban political organizations – stands clearly larger than the others' (Edsall 1984: 235). The dependency of Democrats, furthermore, upon 'big money' contributions rendered many of them also highly vulnerable to direct influence from business interests.

The outcome was predictable enough. When a relatively coherent class force encounters a fragmented opposition which cannot even conceive of its interests in class terms, then the result is hardly in doubt. Institutions like the National Labor Relations Board and OSHA were crippled or turned around to fit business rather than labour agendas. Moody (1988: 120 and Chapter 6) notes, for example, that by 1983 it took on average 627 days for the NLRB to issue a decision on an unfair labour practice: an impossible time to wait if the unfair labour practice involves dismissal and the person dismissed has nothing to live on in the meantime. It was this political and administrative climate of total disregard for laws governing labour rights and occupational health and safety which set the stage for that 'accident waiting to happen' at Hamlet, North Carolina.

The failure to register political anger of the sort that followed the Triangle Shirtwaist Company fire in 1911 in New York City also deserves some comment. A similar event in a relatively remote rural setting posed immediate logistical problems for massive on-the-spot political responses (such as the protest demonstration on Broadway), illustrating the effectiveness of capitalist strategies of geographical dispersal away from politicized central city locations as a means of labour control. The only other path to a generalized political response lay in widespread media attention and public debate: surely, given modern communications technology, a very real possibility. But here the other element to the situation prevailing in 1991 came into play. Not only were the working-class institutions that might have taken up the cause greatly weakened, both in their ability to react as well as in their access to the media, but the very idea of any kind of working-class politics was likewise on the defensive (if not downright discredited in certain 'radical' circles), even though capitalist class interests and the captive Republican party had been waging a no-holds-barred and across-the-board class war against the least privileged sectors of the population for the previous two decades.

This weakening of working-class politics in the United States from the mid-1970s onwards can be tracked back to many causes which cannot be examined in detail here. But one contributory feature has been the increasing fragmentation of 'progressive' politics around special issues and the rise of the so-called new social movements focusing on gender, race, ethnicity, ecology, multiculturalism, community, and the like. These movements often became a working and practical alternative to class politics of the traditional sort and, in some instances, have exhibited downright hostility to such politics.

I think it instructive here to note that, as far as I know, none of the institutions associated with such new social movements saw fit to engage politically with what happened in Hamlet, North Carolina. Women's organizations, for example, were heavily preoccupied with the question of sexual harassment and mobilizing against the Clarence Thomas appointment, even though it was mainly women who died in the North Carolina fire and women who continue to bear an enormous burden of exploitation in the 'Broiler Belt'. And apart from the Rainbow coalition and Jesse Jackson, African-American (and Hispanic) organizations also remained strangely silent on the matter, while some ecologists (particularly the animal rights wing) exhibited more sympathy for the chickens than for the workers. The general tone in the media, therefore, was to sensationalize the horror of the 'accident', but not to probe at all into its origins and certainly not to indite capitalist class interests, the Republican party, the failures of the State of North Carolina, or OSHA as accessory to a murderously negligent event.

DAVID HARVEY

THE POSTMODERN DEATH OF JUSTICE

According to most commonsense meanings of the word, many of us would accept that the conditions under which men, women and minorities work in the Hamlet plant are socially unjust. Yet to make such a statement presupposes that there are some universally agreed norms as to what we mean or ought to mean by the concept of social justice; furthermore it presupposes that no barrier exists, other than the normal ambiguities and fuzziness, to applying the full force of such a powerful principle to the circumstances of North Carolina. But 'universality' is a word which conjures up doubt and suspicion, downright hostility, even, in these 'postmodern' times: the belief that universal truths are both discoverable and applicable as guidelines for political-economic action is nowadays often held to be the chief sin of 'the Enlightenment project' and of the 'totalizing' and 'homogenizing' modernism it supposedly generated.

The effect of the postmodern critique of universalism has been to render any application of the concept of social justice problematic. And there is an obvious sense in which this questioning of the concept is not only proper but imperative – too many colonial peoples have suffered at the hands of Western imperialism's particular justice; too many African-Americans have suffered at the hands of the white man's justice; too many women from the justice imposed by a patriarchal order; and too many workers from the justice imposed by capitalists, to make the concept anything other than problematic. But does this imply that the concept is useless or that to dub events at Hamlet, North Carolina, as 'unjust' has no more force than some localized and contingent complaint?

The difficulty of working with the concept is compounded further by the variety of idealist and philosophical interpretations put upon the term throughout the long history of Western thought on the matter. There are multiple competing theories of social justice and each has its flaws and strengths. Egalitarian views, for example, immediately run into the problem that 'there is nothing more unequal than the equal treatment of unequals' (the modification of doctrines of equality of opportunity in the United States by requirements for affirmative action, for example, have recognized the historical force of that problem). Positive law theories (whatever the law says is just), utilitarian views (the greatest good of the greatest number), social contract and natural right views, together with the various intuitionist, relative deprivation, and other interpretations of justice, all compete for our attention, leaving us with the conundrum: *which* theory of social justice is the most socially just?

Social justice, for all of the universalism to which proponents of a particular version of it might aspire, has long turned out to be a rather heterogeneous set of concepts. To argue for a particular definition of social justice has always implied, therefore, appeal to some higher-order criteria to define which theory of social justice is more appropriate or more just than another. An infinite regress of argument immediately looms as does, in the other direction, the relative ease of deconstruction of any notion of social justice as meaning anything whatsoever, except whatever individuals or groups, given their multiple indentities and functions, at some particular moment find it pragmatically, instrumentally, emotionally, politically, or ideologically useful to mean.

At this point there seem two ways to go with the argument. The first is to look at how the multiple concepts of justice are embedded in language and this leads to theories of meaning of the sort that Wittgenstein advanced and which have had such an important impact upon postmodern ways of thought:

> How many kinds of sentence are there? . . . There are *countless* kinds: countless different kinds of use to what we call 'symbols', 'words', 'sentences'. And this multiplicity is not something fixed, given once for all: but new types of language, new language games, as we may say, come into existence and others become obsolete and get forgotten. . . . Here the term 'language-*game*' is meant to bring into prominence the fact that the *speaking* of language is part of an activity, or a form of life. . . . How did we *learn* the meaning of this word ('good' for instance)? From what sort of examples? In what language games? Then it will be easier for us to see that the word must have a family of meanings.
>
> (Wittgenstein 1967)

From this perspective, social justice has no universally agreed meaning but a 'family' of meanings which can be understood only through the way each is embedded in a particular language game. But we should note two things about Wittgenstein's formulation. First, the appeal to a 'family' of meanings suggests some kind of interrelatedness, and we should presumably pay attention to what those relations might be. Second, each language game attaches to the particular social, communicative, experiential and perceptual world of the speaker. The upshot is to bring us to a point of cultural, linguistic or discourse relativism of some sort, albeit based upon the material circumstances of the subject. We should also, then, pay careful attention to those material circumstances.

The second path is to admit the relativism of discourses about justice, but to insist that discourses are expressions of social power and that the 'family' of meanings derives its interrelatedness precisely through the nature of power relations pertaining within and between different social formations. The simplest version of this idea is to interpret social justice as embedded in the hegemonic discourses of any ruling class or ruling faction. This is an idea which goes back to Plato who, in the *Republic*, has Thrasymachus argue that:

> Each ruling class makes laws that are in its own interest, a democracy democratic laws, a tyranny tyrannical ones and so on; and in making these laws they define as 'right' for their subjects what is in the interest

> of themselves, the rulers, and if anyone breaks their laws he is punished as a 'wrong-doer'. That is what I mean when I say that 'right' is the same in all states, namely the interest of the established ruling class.
>
> (Plato 1965)

Marx and Engels make a similar argument. The latter, for example, writes:

> The stick used to measure what is right and what is not is the most abstract expression of right itself, namely *justice*. . . . The development of right for the jurists . . . is nothing more than a striving to bring human conditions, so far as they are expressed in legal terms, ever closer to the ideal of justice, *eternal* justice. And always this justice is but the ideologized, glorified expression of the existing economic relations, now from their conservative and now from their revolutionary angle. The justice of the Greeks and Romans held slavery to be just; the justice of the bourgeois of 1789 demanded the abolition of feudalism on the ground it was unjust. The conception of eternal justice, therefore, varies not only with time and place, but also with the persons concerned. . . . While in everyday life . . . expressions like right, wrong, justice, and sense of right are accepted without misunderstanding even with reference to social matters, they create . . . the same hopeless confusion in any scientific investigation of economic relations as would be created, for instance, in modern chemistry if the terminology of the phlogiston theory were to retained.
>
> (Marx and Engels 1951: 562–4)

From this it follows that the 'situatedness' or 'standpoint' of whoever makes the argument is relevant if not determinant to understanding the particular meaning put upon the concept. Sentiments of this sort have been taken much further in the postmodern literature. 'Situatedness', 'otherness' and 'positionality' (usually understood in the first instance in terms of class, gender, race, ethnicity, sexual preference, and community, though in some formulations even these categories are viewed with suspicion) here become crucial elements in defining how particular differentiated discourses (be they about social justice or anything else) arise and how such discourses are put to use as part of the play of power. There can be no universal conception of justice to which we can appeal as a normative concept to evaluate some event, such as the Imperial Foods plant fire. There are only particular, competing, fragmented and heterogeneous conceptions of and discourses about justice, which arise out of the particular situations of those involved. The task of deconstruction and of postmodern criticism is to reveal how *all* discourses about social justice hide power relations. The effect of this postmodern extension of Engels' line of reasoning is well described by White. Postmodernists, he says, argue

> that we are far too ready to attach the word 'just' to cognitive, ethical, and political arrangements that are better understood as phenomena of

> power that oppress, neglect, marginalize, and discipline others. In unmasking such claims about justice, postmodern thinkers imply that their work serves some more valid but unspecified notion of justice. One sees this in Derrida's declaration that 'Deconstruction is justice', but also in his cautioning that one can neither speak directly about nor experience justice. In answering the sense of responsibility to otherness, one serves justice but one does so with a sense of the infinite, open-ended character of the task.
>
> (White 1991: 115)

The effect, however, is to produce 'a rather simple bipolar world: deconstructionists and other postmoderns who struggle for justice, and traditional ethical and political theorists who are the ideologues of unjust orders'. And this, in turn, produces a serious dilemma for all forms of postmodern argumentation:

> On the one hand, its epistemological project is to deflate all totalistic, universalistic efforts to theorize about justice and the good life; and yet on the other hand, its practical project is to generate effective resistance to the present dangers of totalizing, universalizing rationalization processes in society. In short, the source of much injustice in contemporary society is seen as general and systematic; the response, however, bars itself from normatively confronting the problem on a comparable level by employing a theory of justice offering universally valid, substantive principles. Postmodern reflection thus seems to deny itself just the sort of normative armament capable of conducting a successful fight.
>
> (White 1991: 116)

We can see precisely this difficulty emerging in the circumstances that led up to events in Hamlet, North Carolina. When business organized itself as a class to attack government regulation and intervention and the welfare state (with its dominant notions of social rationality and just redistributions), it did so in the name of the unjust and unfair regulation of private property rights and the unfair taxation of the proper fruits of entrepreneurial endeavour in freely functioning markets. Just deserts, it has long been argued by the ideologues of free-market capitalism (from Adam Smith onwards), are best arrived at through competitively organized, price-fixing markets in which entrepreneurs are entitled to hang on to the profit engendered by their endeavours. There is then no need for explicit theoretical, political or social argument over what is or is not socially just because social justice is whatever is delivered by the market. Each 'factor' of production (land, labour and capital), for example, will receive its marginal rate of return, it just reward, according to its contribution to production. The role of government should be confined to making sure that markets function freely (e.g. by curbing monopoly powers) and that they are 'properly organized' (which may extend to compensating for clear cases of market failure in, for example, the case of unpriced externalities such as environmental pollution and health hazards).

It does not, of course, take that much sophistication to deconstruct this conception of justice as a manifestation of a particular kind of political-economic power. Yet there is widespread, perhaps even hegemonic acceptance of such a standpoint as the numerous 'tax revolts' in the United States over the last decades have shown. From this standpoint, the incident in North Carolina can be interpreted as an unfortunate accident, perhaps compounded by managerial error, in a basically just system which (a) provides employment where there otherwise would be none at wages determined by the demand and supply conditions prevailing in the local labour market, and (b) fills the shops (contrast the ex-Soviet Union) with a vast supply of cheap protein which poor people can for the most part afford to buy. In so far as this doctrine of just deserts in the marketplace is ideologically hegemonic, protest in the North Carolina case would be minimized and confined simply to an enquiry into who it was that locked the doors. The lack of response to the Hamlet case can therefore be interpreted as an indication of precisely how dominant such a notice of justice is in the United States today.

The obvious discourse with which to confront such arguments resides in doctrines of workers' rights and the whole rhetoric of class struggle against exploitation, profit-making and worker disempowerment. Neither Marx nor Engels would here eschew *all* talk of rights and justice. While they clearly recognize that these concepts take on different meanings across space and time and according to persons, the exigencies of class relations inevitably produce, as Marx (1967: 235) argues in the case of the fight between capital and labour over the proper length of the working day, 'an antinomy, right against right, both equally bearing the seal of the law of exchanges'. Between such *equal* rights (that of the capitalist and that of the worker) 'force decides'. What is at stake here is not the arbitration between competing claims according to some universal principle of justice, but class struggle over the particular conception of justice and rights which shall be applied to a given situation. In the North Carolina case, had the rights of workers to be treated with respect under conditions of reasonable economic security and safety and with adequate remuneration been properly respected, then the incident almost certainly would not have happened. And if all workers (together with the unemployed) were accorded the same rights and if the exorbitant rates of profit in broiler chicken processing (as well as in other industries) had been curbed, then the importance of the relatively low price of this source of protein for the poor would have been significantly diminished.

The problem, however, is that such working-class rhetoric on rights and justice is as open to criticism and deconstruction as its capitalistic equivalent. Concentration on class alone is seen to hide, marginalize, disempower, repress and perhaps even oppress all kinds of 'others' precisely because it can not and does not acknowledge explicitly the existence of heterogeneities and differences based on, for example, race, gender, sexuality, age, ability, culture, locality, ethnicity, religion, community, consumer preferences, group affiliation, and the like. Open-ended responsibility to all of these multiple othernesses makes it difficult if not impossible to respond to events in North Carolina with a single institutionalized discourse which might be maximally effective in confronting the rough justice of capitalism's political economy at work in the Broiler Belt.

We here encounter a situation with respect to discourses about social justice which closely matches the political paralysis exhibited in the failure to respond to the North Carolina fire. Politics and discourses both seem to have become so mutually fragmented that response is inhibited. The upshot appears to be a double injustice: not only do men and women, whites and African-Americans die in a preventable event, but we are simultaneously deprived of any normative principles of justice whatsoever by which to condemn or indite the responsible parties.

THE RESURRECTION OF SOCIAL JUSTICE

There are abundant signs of discontent with the impasse into which post-modernism's and post-structuralism's approach to the question of social justice has fallen. And a number of different strategies have emerged to try to resurrect the mobilizing power of arguments about justice in ways which either permit appeal to carefully circumscribed but nevertheless general principles or which, more ambitiously, try to build a bridge between the supposed universalisms of modernism and the fragmented particularities left behind by post-structuralist deconstructions. I note, for example, Walzer's (1983) attempt to pluralize the concept of justice as equality so as to respect the cultural creations of others, and Peffer's (1990) attempt to construct principles of social justice that are consistent with Marxist social theory as an antidote to that wing of Marxism which regards all talk of justice and of rights as a pernicious bourgeois trap. From multiple directions, then, there emerges a strong concern to reinstate concern for social justice and to re-elaborate upon what it takes to create the values and institutions of a reasonably just society.

I think it important at the outset to concede the seriousness of the radical intent of post-structuralists to 'do justice' in a world of infinite heterogeneity and open-endedness. Their reasons for refusing to apply universal principles rigidily across heterogeneous situations are not without considerable weight. This alerts us to the unfortunate ways in which many social movements in the twentieth century have foundered on the belief that because their cause is just they cannot possibly themselves behave unjustly. The warning goes even deeper: the application of *any* universal principle of social justice across heterogeneous situations is certain to entail some injustice to someone, somewhere. But, on the other hand, at the end of a road of infinite heterogeneity and open-endedness about what justice might mean, there lies at best a void or at worst a rather ugly world in which the needs of rapists (a particular form of 'otherness' after all) are 'negotiated' or even regarded as 'just' on equivalent terms with those of their victims. Affirming the importance of infinite heterogeneity and open-endedness directly connects to the charge against post-structuralism that it is an 'anything goes' way of thinking within which no particular moral or ethical principles can carry any particular weight over any other. 'At some point', says White (1991: 133), 'one must have a way of arguing that not all manifestations of otherness should be fostered; some ought to be constrained.' And this presumes some general principles of right or justice.

There is, White goes on to assert, often a tacit admission of such a problem in some of post-structuralism's founding texts. Foucault (1980: 107–8), having argued strenuously that we can never disentangle 'mechanisms of discipline' from principles of right, ends up raising the possibility of 'a new form of right, one which must be anti-disciplinarian, but, at the same time liberated from the principle of sovereignty'. Lyotard likewise argues explicitly for the creation of a 'pristine' but 'non-consensual' notion of justice in *The Postmodern Condition*. And Derrida is deeply concerned about ethics. But in no case are we told much about what, for example, a 'new form of right' might mean.

Initiatives have consequently emerged to try to resurrect some general principles of social justice while attending to post-structuralist criticisms of universalizing theory which marginalizes 'others'. There are two particular lines of argument that appear to be potentially fruitful:

1 breaking out of the local;
2 situating 'situated knowledges'.

Breaking out of the local

The first line derives from the observation that most post-structuralist critical interventions tend to confine their radicalizing politics to social interactions occurring 'below the threshold where the systemic imperatives of power and money become so dominant' (White 1991: 107). The politics of resistance which they indicate are typically attached to small-scale communities of resistance, marginalized groups, abnormal discourses, or simply to that zone of personal life sometimes termed 'the life world', which can be identified as distinct from and potentially resistant to penetration by the rationalizing, commodified, technocratic and hence alienating organization of contemporary capitalism. It is hard to read this literature without concluding that the objective of reform or revolutionary transformation of contemporary capitalism as a whole has been given up on, even as a topic for discussion, let alone as a focus for political organization. This 'opting out' from consideration of a whole range of questions is perhaps best signalled by the marked silence of most postmodern and post-structuralist thinkers when it comes to critical discussion of any kind of political economy, let alone that of the Marxian variety. The best that can be hoped for, as someone like Foucault seems to suggest, is that innumerable localized struggles might have some sort of collective effect on how capitalism works in general.

Dissatisfaction with such a politics has led some socialist feminists in particular (see Fraser 1989; Young 1990a, 1990b) to seek ways to broaden the terrain of struggle beyond the world of face-to-face communalism and into battles over such matters as welfare state policy, public affairs, political organization via, in Fraser's case, 'an ethic of solidarity' and, in Young's case, through explicit statement of norms of social justice.

Young (1990a: 300–2), for example, complains that the attempt to counter 'the alienation and individualism we find hegemonic in capitalist patriarchal society', has led feminist groups 'impelled by a desire for closeness and mutual identification', to construct an ideal of community 'which generates borders, dichotomies, and exclusions' at the same time as it homogenizes and represses difference within the group. She explicitly turns the tools of deconstruction against such ideals of community in order to show their oppressive qualities:

> Racism, ethnic chauvinism, and class devaluation, I suggest, grow partly from a desire for community, that is from the desire to understand others as they understand themselves and from the desire to be understood as I understand myself. Practically speaking, such mutual understanding can be approximated only within a homogeneous group that defines itself by common attributes. Such common identification, however, entails reference also to those excluded. In the dynamics of racism and ethnic chauvinism in the United States today, the positive identification of some groups is often achieved by first defining other groups as the other, the devalued semihuman.
>
> (Young 1990a: 311–12)

Young, however, 'parts ways' with Derrida because she thinks it 'both possible and necessary to pose alternative conceptualizations' (Young 1990a: 321). The first step to her argument is to insist that individuals be understood as 'heterogeneous and decentered' (see below). No social group can be truly unitary in the sense of having members who hold to a singular identity. Young strives on this basis to construct some norms of behaviour in the public realm. Our conception of social justice 'requires not the melting away of differences, but institutions that promote reproduction of and respect for group differences without oppression' (Young 1990b: 47). We must reject 'the concept of universality as embodied in republican versions of Enlightenment reason' precisely because it sought to 'suppress the popular and linguistic heterogeneity of the urban public' (ibid.: 108). 'In open and accessible public spaces and forums, one should expect to encounter and hear from those who are different, whose social perspectives, experience and affiliations are different.'

The ideal to which she appeals is 'openness to unassimilated otherness'. This entails the celebration of the distinctive cultures and characteristics of different groups and of the diverse group identities which are themselves perpetually being constructed and deconstructed out of the flows and shifts of social life. But we here encounter a major problem. In modern mass urban society, the multiple mediated relations that constitute that society across time and space are just as important and as 'authentic' as unmediated face-to-face relations. It is just as important for a politically responsible person to know about and respond politically to all those people who daily put breakfast upon our table, even though market exchange hides from us the conditions of life of the producers (see Harvey 1990). When we eat chicken, we relate to workers we never see of the sort that died in Hamlet, North Carolina. Relationships between individuals get mediated through market functions and state powers, and we have to define conceptions of justice capable of operating across and through these multiple mediations. But this is the realm of politics which postmodernism typically avoids.

Young here proposes 'a family of concepts and conditions' relevant to a contemporary conception of social justice. She identifies 'five faces of oppression' which are:

1 *exploitation*: the transfer of the fruits of the labour from one group to another, as, for example, in the cases of workers giving up surplus value to capitalists or women in the domestic sphere transferring the fruits of their labour to men;
2 *marginalization*: the expulsion of people from useful participation in social life so that they are 'potentially subjected to severe material deprivation and even extermination';
3 *powerlessness*: the lack of that 'authority, status, and sense of self' which would permit a person to be listened to with respect;
4 *cultural imperialism*: stereotyping in behaviours as well as in various forms of cultural expression such that 'the oppressed group's own experience and interpretation of social life finds little expression that touches the dominant culture, while that same culture imposes on the oppressed group its experience and interpretation of social life';
5 *violence*: the fear and actuality of random, unprovoked attacks, which have 'no motive except to damage, humiliate, or destroy the person'.

I would want to add a further dimension concerning freedom from the oppressive *ecological consequences* of others' actions.

This multi-dimensional conception of social justice is extremely useful. It alerts us to the existence of a 'long social and political frontier' of political action to roll back multiple oppressions. It also emphasizes the heterogeneity of experience of injustice – someone unjustly treated in the workplace can act oppressively in the domestic sphere, and the victim of that may, in turn, resort to cultural imperialism against others. Yet there are many situations, such as those in Hamlet, North Carolina, where multiple forms of oppression coalesce. Young's conception of a just society combines, therefore, the requirement of freedom from these different forms of oppression (occurring in mediated as well as in face-to-face situations) with 'openness to unassimilated otherness'. However,

> [T]he danger in affirming difference is that the implementation of group-conscious policies will reinstate stigma and exclusion. In the past, group-conscious policies were used to separate those defined as different and exclude them from access to the rights and privileges enjoyed by dominant groups. . . . Group-conscious policies cannot be used to justify exclusion of or discrimination against members of a group in the exercise of general political and civil rights. A democratic cultural pluralism thus requires a dual system of rights: a more general system of rights which are the same for all and a more specific system of group-conscious policies and rights.
>
> (Young 1990b: 174)

The double meaning of universality then becomes plain: 'universality in the sense of the participation and inclusion of everyone in moral and social life does not imply universality in the sense of adoption of a general point of view that leaves behind particular affiliations, feelings, commitments, and desires' (ibid.: 105). Universality is no longer rejected out of hand, but reinserted in a dialectical relation to particularity, positionality and group difference. But what constitutes this universality?

Situating 'situated knowledges'

The second line of development derives from reflection on what it means to say that all knowledge (including conceptions of social justice and of social needs) are 'situated' in a heterogeneous world of difference. 'Situatedness' can be construed, however, in different ways. What I shall term the 'vulgar' conception of it dwells almost entirely on the relevance of individual biographies: I see, interpret, represent and understand the world in the way I do because of the particularities of my life history. The separateness of language games and discourses is emphasized, and difference is treated as biograpically and sometimes even institutionally, socially, historically and geographically determined. It proceeds as if none of us can throw off even some of the shackles of personal history or internalize what the condition of being 'the other' is all about and leads to an exclusionary politics of the sort that Young rejects. And it is frequently used as a rhetorical device either to enhance the supposed authenticity and moral authority of one's own accounts of the world or to deny the veracity of other accounts ('since she is black and female of rural origins she cannot possibly have anything authentic to say about conditions of life of the white bourgeoisie in New York City' or, more commonly, 'Because he is white, male, Western, heterosexual he is bound to be tied to a certain vision of how the world works'). Individual biographies do, of course, matter: all sorts of problems arise when someone privileged (like myself) purports to speak for or even about others. This is a difficult issue for contemporary social science and philosophy to confront, as Spivak (1988) shows. But a relativist, essentialist and non-dialectical view of situatedness generates immense political difficulties. I would not be permitted to speak about the experiential horror of the North Carolina fire, for example, because I am not working class, nor a woman, nor an African-American (nor, for that matter, was I killed in it). Economically secure, professional, white feminists could not, likewise, speak for any woman whose situation is different. No one, in fact, could assume the right or obligation to speak for 'others', let alone against the oppression of anyone whose identity is construed as 'other'.

There is, however, a far profounder and more dialectical sense of 'situatedness' to which we can appeal. In Hegel's parable of the master and the slave, for example, situatedness is not seen as *separate and unrelated* difference, but as a *dialectical power relation* between the oppressed and the oppressor. Marx appropriated and radically transformed the Hegelian dialectic in his examination of the relation between capital and labour; his long and critical engagement with bourgeois philosophy and political economy then became the means to define an alternative subaltern and subversive science situated from the perspective of the proletariat. Feminist writers such as Haraway (1990) and Hartsock (1987) examine gender difference and ground their feminist theory in a similar way.

Such a dialectical conception pervades Derrida's view of the individual subject as someone who has no solid identity, but who is a bundle of heterogeneous and not necessarily coherent impulses and desires. Multiple forms of interaction with the world construct individuals as 'a play of difference that cannot completely be comprehended' (Young 1990: 232). 'Otherness' is thereby necessarily internalized within the self. 'Situatedness' is then taken out of its wooden attachment to identifiable individuals and their biographies and is itself situated as a play of difference. When I eat Kentucky-fried chicken, I am situated at one point in a chain of commodity production that leads right back to Hamlet, North Carolina. When I interact with my daughter, I am inevitably caught in a game of the construction of gender identities. When I refrain from using bait to destroy the slugs that have eaten every flower I have nurtured, then I situate myself in an ecological chain of existence. Individuals are heterogeneously constructed subjects, internalizing 'otherness' by virtue of their relations to the world. Spivak's (1988: 294–308) answer to the whole dilemma of political representation of the other then rests on invoking Derrida's call to render 'delirious that interior voice that is the voice of the other in us'.

Unfortunately, this does not exhaust the problem for, as Ricoeur (1991) notes, our own sense of selfhood and of identity in part gets constructed through the narrative devices that we use to describe our temporal relation to the world, and so assumes relatively durable configurations. While identity does not rest upon sameness or essence, it does acquire durability and permanence according to the stories we tell ourselves and others about our history. Although identity internalizes otherness, it nevertheless delimits and renders relatively durable both the field of 'othernesses' brought into play and the relation of those others to a particular sense of selfhood. Whites may construct their identity through historical development of a particular relation to blacks, for example; indeed, both groups may use the other to construct themselves. This intertwining of black and white identities in American history was, as Gates (1992) has recently shown, fundamental to James Baldwin's conception of race relations. But it is precisely by such means that much of the racial problematic of contemporary culture resides.

Nevertheless, we can, from this dialectical perspective, better appreciate Hartsock's (1987) claim that 'attention to the epistemologies of situated knowledges', can 'expose and clarify the theoretical bases for political alliance and solidarity' at the same time as it provides 'important alternatives to the dead-end oppositions set up by postmodernism's rejection of the Enlightenment'. We must pay close attention to the 'similarities that can provide the basis for differing groups to understand each other and form alliances'. Refusing the postmodern formulation of the problem, Hartsock insists that we engage with dominant discourses precisely because we cannot abstract from the complex play of power relations. That wing of postmodernism that holds to the 'vulgar' version of situatedness cannot engage with the dominant lines of political-economic power at work under capitalism, and thereby typically marginalizes itself. This parallels my own conclusion in *The Condition of Postmodernity*:

> while [postmodernism] opens up a radical prospect by acknowledging the authenticity of other voices, postmodernist thinking immediately shuts off those other voices from access to more universal sources of power by ghettoizing them within an opaque otherness, the specificity of this or that language game. It thereby disempowers those voices (of women, ethnic and racial minorities, colonized peoples, the unemployed, youth, etc.) in a world of lop-sided power relations. The language game of a cabal of international bankers may be impenetrable to us, but that does not put it on a par with the equally impenetrable language of inner-city blacks from the standpoint of power relations.
>
> (Harvey 1989: 117)

By insisting upon mutually exclusionary discourses of the sort to which the narrow definition of situatedness gives rise, we would foreclose upon the most obvious implication of the North Carolina fire: that pursuit of working-class politics might protect, rather than oppress and marginalize, interests based on gender and race even if that working-class politics regrettably makes no explicit acknowledgement of the importance of race and gender. The failure of a feminist movement strongly implanted within the professions in the United States to respond to events in North Carolina while mobilizing around the nomination of a Supreme Court judge, either suggests that narrowly construed views of situatedness have rather more practical political purchase than many would care to admit, or else it tacitly orders situatedness in such a way that what happened to those 'others' in North Carolina was viewed as somehow less important than the nomination of a Supreme Court judge of highly dubious moral standing. They were not necessarily wrong in this, for as Haraway points out, it is not *difference* which matters, but *significant* difference:

> In the consciousness of our failures, we risk lapsing into boundless difference and giving up on the confusing task of making partial, real connection. Some differences are playful, some are poles of world historical systems of domination. Epistemology is about knowing the difference.
>
> (Haraway 1990: 202–3)

But what is this 'epistemology' which permits us to know the difference? How should we pursue it? And to what politics does it give rise?

CLASS RELATIONS, SOCIAL JUSTICE AND THE POLITICS OF DIFFERENCE

There are a number of disparate threads to be drawn together in the guise of a general conclusion. On the one hand, we find a line of argument about social justice that passes through postmodernism and post-structuralism to arrive at a point of recognition that some kinds of (unspecified) universals are necessary and that some sort of epistemology (unspecified) is needed to establish when, how and where difference and heterogeneity are significant. On the other hand, we have a political economic situation, as characterized by the North Carolina deaths, which indicates a seeming paralysis of progressive politics in the face of class oppression. How, then, are we to link the two ends of this theoretical and political tension?

Consider, first, the obvious lesson of the Imperial Foods plant fire: that an effective working-class politics would have better protected the rights of men and women, whites and African-Americans in a situation where those particular identities, rather than those of class, were not of primary significance. This conclusion merits embellishment and I will look at it primarily in relation to feminist politics. Lynn Segal (1991) has recently noted that 'despite the existence of the largest, most influential and vociferous feminist movement in the world, it is US women who have seen the least *overall* change in the relative disadvantages of their sex, compared to other Western democracies' over the past 20 years. The huge gains made in the United States by women within 'the most prestigious and lucrative professions' have been offset entirely by a life of increasing frustration, impoverishment and powerlessness for the rest. The feminization of poverty (not foreign to Hamlet, North Carolina) has been, for example, one of the most startling social shifts in the United States over the past two decades, a direct casualty of the Republican party class war against the welfare state and working-class rights and interests. 'In countries where there have been longer periods of social-democratic government and stronger trade unions,' Segal continues, 'there is far less pay-differential and occupational segregation (both vertical and horizontal) between women and men, and far greater expansion of welfare services.' Given the far superior material conditions of life achieved for women in such social democracies (and I also note parenthetically the savage diminution in many women's rights since 1989 in what was once the communist bloc), 'it seems strange for feminists to ignore the traditional objectives of socialist or social-democratic parties and organised labour', even though such institutions have obvious weaknesses and limitations as vehicles for pursuit of feminism's objectives (see, for example, Fraser's (1989) compelling argument concerning the gender bias implicit in many welfare state policies). Nevertheless, Segal continues, 'at a time when the advances made by some women are so clearly overshadowed by the increasing poverty experienced so acutely by others (alongside the unemployment of the men of their class and group), it seems perverse to pose women's specific interests *against* rather than *alongside* more traditional socialist goals.' Unless, of course, 'women's interests' are either construed in a very narrow professional and class-biased sense or seen as part of 'an endless game of self-explorations played out on the great board of Identity' (Segal 1991: 90–1).

Segal here parallels Hartsock's concern for the 'bases for political alliances and solidarity'. This requires that we identify 'the *similarities* that can provide the basis for differing groups to understand each other and form alliances'. Young likewise ties the universality criteria she deploys to the idea that 'similarity is never sameness'. Difference can never be characterized, therefore, as 'absolute otherness, a complete absence of relationship or shared attributes'. The *similarity* deployed to measure *difference and otherness* requires, then, just as close an examination (theoretically as well as politically) as does the production of otherness and difference itself. Neither can be established without the other. To discover the basis of similarity (rather than to presume sameness) is to uncover the basis for alliance formation between seemingly disparate groups.

But in today's world, similarity largely resides in that realm of political-economic action so often marginalized in post-structuralist accounts, for it is in terms of commodities, money, market exchange, capital accumulation, and the like that we find ourselves sharing a world of similarity increasingly also characterized by homogeneity and sameness. The radical post-structuralist revolt against that sameness (and its mirror image in some forms of working-class politics) has set the tone of recent debates. But the effect has been to throw out the living baby of political and ethical solidarities and similarities across differences, with the cold bathwater of capitalist-imposed conceptions of universality and sameness. Only through critical re-engagement with political economy, therefore, can we hope to re-establish a conception of social justice as something to be fought for as a key value within an ethics of political solidarity.

Although the conception of justice varies 'not only with time and place, but also with the persons concerned', we must also here recognize the political force of the fact that a particular conception of it can be 'accepted without misunderstanding' in everyday life. Though 'hopelessly confused' when examined in abstraction, ideals of social justice can still function (as Engels' example of the French Revolution allows) as a powerful mobilizing discourse for political action.

But two decades of postmodernism and post-structuralism have left us with little basis to accept any particular norm of social justice 'without misunderstanding', while in everyday life a titanic effort unfolds to convince all and sundry that any kind of regulation of market freedoms or any level of taxation is unjust. Empowerment is then conceived of (as none other than John Major now avows through his active use of the term) as leaving as much money as possible in the wage-earners' as well as in the capitalists' pockets; freedom and justice are attached to maximizing market choice; and rights are interpreted as a matter of consumer sovereignty free of any government dictates. Perhaps the most important thing missing from the postmodern debate these last two decades is the way in which this right-wing and reactionary definition of market justice and of rights has played such a revolutionary role in creating the kind of political economy which produced the effects of the North Carolina fire.

Under such circumstances, reclaiming the terrain of justice and of rights for progressive political purposes appears as an urgent theoretical and political task. But in order to do this, we have to come back to that 'epistemology' which helps us tell the difference between significant and non-significant others, differences and situatedness, and which will help promote alliance formation on the basis of similarity rather than sameness. My own epistemology for this purpose rests on a modernized version of historical and geographical materialism, which forms a meta-theoretical framework for examining not only how differences understood as power relations are produced through social action, but also how they acquire the particular significance they do in certain situations. From this standpoint, it is perfectly reasonable to hold on the one hand that the philosophical, linguistic and logical critiques of universal propositions about social justice are correct, while acknowledging on the other hand the putative power of appeals to social justice in certain situations, such as the contemporary United States, as a basis for political action. Struggles to bring a particular kind of discourse about justice into a hegemonic position have then to be seen as part of a broader struggle over ideological hegemony between conflicting groups in any society.

CONCLUSIONS

The overall effect is to leave us with some important analytical, theoretical and political tasks, which can be summarized as follows:

First, the universality condition can never be avoided, and those who seek so to do (as is the case in many postmodern and post-structuralist formulations) only end up hiding rather than eliminating the condition. But universality must be construed in dialectical relation with particularity. Each defines the other in such a way as to make the universality criterion always open to negotiation through the particularities of difference. It is useful here to examine the political-economic processes by which society actually achieves such a dialectical unity. Money, for example, possesses universal properties as a measure of value and medium of exchange at the same time as it permits a wide range of highly decentralized and particularistic decision-making in the realm of market behaviours which feed back to define what the universality of money is all about. It is precisely this dialectic which gives strength to right-wing claims concerning individual freedoms and just deserts through market co-ordinations. While the injustice that derives is plain – the individual appropriation and accumulation of the social power which money represents produces massive and ever-widening social inequality – the subtle power of the universality–particularity dialectic at work in the case of money has to be appreciated. The task of progressive politics is to find an equally powerful, dynamic and persuasive way of relating the universal and particular in the drive to define social justice from the standpoint of the oppressed.

Second, respect for identity and 'otherness' must be tempered by the recognition that although all others may be others, 'some are more other than others', and that in any society certain principles of exclusion have to operate. How this exclusion shall be gauged is embedded in the first instance in a universality condition which prevents groups from imposing their will oppressively on others. This universality condition cannot, however, be imposed hierarchically from above: it must be open to constant negotiation, precisely because of the way in which disparate claims may be framed (for example, when the rich demand that the oppressive sight (to them) of homelessness be cleared from their vision by expelling the homeless from public spaces).

Third, all propositions for social action (or conceptions of social justice) must be critically evaluated in terms of the situatedness or positionality of the argument and the arguer. But it is equally important to recognize that the individuals developing such situated knowledge are not themselves homogeneous entities but bundles of heterogeneous impulses, many of which derive from an internalization of 'the other' within the self. Such a conception of the subject renders situatedness itself heterogeneous and differentiated. In the last instance, it is the social construction of situatedness which matters.

Fourth, 'epistemology that can tell the difference' between significant and insignificant differences or 'othernesses' is one which can understand the social processes of construction of situatedness, otherness, difference, political identity and the like. And we here arrive at what seems to me to be the most important epistemological point: the relation between social processes of construction of identities on the one hand and the conditions of identity politics on the other. If respect for the condition of the homeless (or the racially or sexually oppressed) does not imply respect for the social processes creating homelessness (or racial or sexual oppression), then identity politics must operate at a dual level. A politics which seeks to eliminate the processes which give rise to a problem looks very different from a politics which merely seeks to give full play to differentiated identities once these have arisen.

We encounter, here, a peculiar tension. The identity of the homeless person (or the racially oppressed) is vital to their sense of selfhood. Perpetuation of that sense of self and of identity may depend on perpetuation of the processes which gave rise to it. A political programme that successfully combats homelessness (or racism) has to face up to the real difficulty of a loss of identity on the part of those who have been victims of such forms of oppression. And there are subtle ways in which identity, once acquired, can precisely by virtue of its relative durability seek out the social conditions (including the oppressions) necessary for its own sustenance.

It then follows that the mere pursuit of identity politics as an end in itself (rather than as a fundamental struggle to break with an identity which internalizes oppression) may serve to perpetuate rather than to challenge the persistence of those processes which gave rise to those identities in the first place. This is a pervasive problem even within the ideological debates swirling around identity politics in academia. And it is a problem which is not new, for as Spivak notes of the French post-structuralists:

> [They] forget at their peril that [their] whole overdetermined exercise was in the interest of a dynamic economic situation requiring that interests, motives (desires), and power (of knowledge) be ruthlessly dislocated. To invoke that dislocation now as a radical discovery that should make us diagnose the economic . . . as a piece of dated analytic machinery may well be to continue the work of that dislocation and unwittingly to help in securing 'a new balance of hegemonic relations'.
> (Spivak 1988: 280)

Perhaps this is the best of all possible lessons we can learn from the political failure to respond to events in Hamlet, North Carolina, and from the lack of any convincing discourse about social justice with which to confront it. For if the historical and geographical *process* of class war waged by the Republican party and the capitalist class these last few years in the United States has feminized poverty, accelerated racial oppression, and further degraded the ecological conditions of life, then it seems that a far more united politics can flow from a determination to check *that* process than will likely flow from an identity politics which largely reflects its fragmented results.

POSTSCRIPT

In September 1992, almost a year to the day after the Imperial Foods plant fire, the owner, Emmett Roe, was sentenced (after an extraordinary plea-bargaining arrangement in which he pleaded guilty to 25 counts of involuntary manslaughter) to 19 years and 11 months in prison. The plea-bargaining arrangement allowed the two managers of the plant, including his son, to go free. Since Emmett Row was 73 years old, he will almost certainly be paroled on health and/or compassionate grounds within a few years. A few days after the Clinton election victory – and the timing is probably significant – three insurance companies that had been resisting payment of any insurance claims (on the grounds that the safety conditions at the plant were so bad that the now-bankrupt Imperial Foods company itself was legally liable, not the insurance companies) finally agreed to pay out $16.1 million to the 101 families that had filed claims for dead, injured or emotionally disturbed workers. Suits against the United States Department of Agriculture are still pending.

REFERENCES

Davidson, O. G. (1991) 'It's still 1911 in America's rural sweatshops', *Baltimore Sun*, 7 Sept., 7A.

Edsall, T. (1984) *The New Politics of Inequality*, New York: Norton.

Foucault, M. (1980) *Power/Knowledge*, London: Harvester-Wheatsheaf.

Fraser, N. (1989) *Unruly Practices*, Minneapolis: University of Minnesota Press.

Gates, H. L. (1992) 'The welcome table: remembering James Baldwin', paper delivered to *Wissenshaftliche Jahrestagung der Deutschen Gesellschaft für Amerikastudien*, Berlin, June.

Haraway, D. (1990) 'A manifesto for cyborgs: science, technology, and socialist feminism in the 1980s', in L. Nicholson (ed.) *Feminism/Postmodernism*, London: Routledge.

Hartsock, N. (1987) 'Rethinking modernism: minority versus majority theories', *Cultural Critique* 7: 187–206.

Harvey, D. (1989) *The Condition of Postmodernity*, Oxford: Basil Blackwell.

——— (1990) 'Between space and time: reflections on the geographical imagination', *Annals, Association of American Geographers* 80: 418–34.

Lyotard, J.-F. (1984) *The Postmodern Condition*, Manchester: University of Manchester Press.

Marx, K. (1967) *Capital*, vol. 1, New York: International Publishers.

Marx, K. and Engels, F. (1951) *Selected Works*, vol. 1, Moscow: Progress Publishers.

Moody, K. (1988) *An Injury to All*, London: Verso.

Peffer, R. (1990) *Marxism, Morality, and Social Justice*, Princeton, NJ: Princeton University Press.

Plato (1965) *The Republic*, Harmondsworth, Middlesex: Penguin Books.

Ricoeur, P. (1991) 'Narrative identity', in D. Wood (ed.) *On Paul Ricoeur: Narrative and Interpretation*, London: Routledge.

Segal, L. (1991) 'Whose left: socialism, feminism and the future', *New Left Review* 185: 81–91.

Spivak, G. (1988) 'Can the subaltern speak?', in C. Nelson and L. Grossberg (eds)

Marxism and the Interpretation of Culture, Urbana, Ill.: University of Illinois Press.
Struck, D. (1991) 'South's poultry plants thrive, feeding on workers' need', *Baltimore Sun*, 8 Sept., Section A.
Walzer, M. (1983) *Spheres of Justice: A Defense of Pluralism and Equality*, Oxford: Basil Blackwell.
White, S. (1991) *Political Theory and Postmodernism*, Cambridge: Cambridge University Press.
Wittgenstein, L. (1967) *Philosophical Investigations*, Oxford: Basil Blackwell.
Young, I. M. (1990a) 'The ideal of community and the politics of difference', in L. Nicholson (ed.) *Feminism/Postmodernism*, London: Routledge.
—— (1990b) *Justice and the Politics of Difference*, Princeton, NJ: Princeton University Press.

4

GROUNDING METAPHOR

Towards a spatialized politics

Neil Smith and Cindi Katz

> Metaphor and metonymy, then. These familiar concepts are borrowed, of course, from linguistics. Inasmuch, however, as we are concerned not with words but rather with space and spatial practice, such conceptual borrowing has to be underwritten by a careful examination of the relationship between space and language.
>
> (Henri Lefebvre)

THE REASSERTION OF SPACE

With the reassertion of space in social and cultural theory, an entire spatial language has emerged for comprehending the contours of social reality.[1] A response in part to the widespread historicism that has dominated 'Western' social thought over the last century and a half, this resurgence of interest in space and spatial concepts is broad based. It was the explicit goal of critical geographic and political economic theory from the late 1960s onwards, a central component of structural and post-structural social analyses, and a core concern of information theory. Most recently, space has provided an attractive lexicon for many feminist, postmodernist, and postcolonial enquiries, the focus for public art and geo art, and a grammar in cultural discourse more broadly.[2] The language of social and cultural investigation is increasingly suffused with spatial concepts in a way that would have been unimaginable two decades ago.

The extent of this reassertion of space is in some respects astonishing. One has to look back to the *fin de siècle* to find an equivalent period in which space was comparably 'on the agenda'. But for all the reconfiguration of spatial and temporal concepts that accompanied the rise of literary and artistic modernism, the privileging of time over space was not so explicitly questioned during this period.[3] The politics of space was not as evident then either, even if the reconceptualization of space around the turn of the last century actually expressed very deep-seated political shifts in economy and culture. Today, from post-Fordism to postmodernism, equally significant shifts are afoot, but unlike a century ago, the current interest in space has provoked diverse appeals for an explicitly 'spatial politics'. If these appeals came first from social theorists like Henri Lefebvre and David Harvey, they may have found their most extreme expression elsewhere; for example, in the work of the literary critic, Fredric Jameson, who claims that contemporary culture is 'increasingly dominated by space and spatial logic', and who concludes that 'a model of political culture appropriate for our own situation will necessarily have to raise spatial issues as its fundamental organizing concern' (Jameson 1984: 71, 89).[4] Jameson conceived this spatial politics as guided by the process he called 'cognitive mapping', a label which, he later declared, was a 'metaphor for class struggle'.[5]

The breadth of interest in space is matched by the breadth of spatial concepts newly in vogue. In social theory and literary criticism, spatial metaphors have become a predominant means by which social life is understood. 'Theoretical spaces' have been 'explored', 'mapped', 'charted', 'contested', 'colonized', 'decolonized', and everyone seems to be 'travelling'. But, perhaps surprisingly, there has been little, if any, attempt to examine the different implications of material and metaphorical space. Metaphorical concepts and uses of 'space' have evolved quite independently from materialist treatments of space, and many of the latter are cast in ways that suggest equal ignorance of the productive entailments of spatial metaphors. Yet if a new spatialized politics is to be both coherent and effective, it will be necessary to comprehend the interconnectedness of material and metaphorical space. In this chapter we argue that many current spatial metaphors, such as 'positionality', 'locality', 'grounding', 'displacement', 'territory', 'nomadism', and so forth require urgent critical scrutiny. The appeal of these spatial metaphors lies precisely in the new meanings they impart, but it is increasingly evident that these metaphors depend overwhelmingly on a very specific and contested conception of space and that they embody often unintended political consequences. At the very least, it is necessary to devise more explicit translation rules, or certainly a critical awareness of the translations connecting material and metaphorical space.

Lest we be misunderstood, this is in no way an argument against metaphor or against spatial metaphor in particular. Metaphor is inseparable from the generation of meaning, from language and thought. Any project to abolish metaphor is not only doomed to failure but is, literally, absurd. Equally, it is not an attempt somehow to separate material and metaphorical space in an unrealistic dualism; rather the lack of enquiry concerning the implications of spatial metaphors suggests an undifferentiated fusion of material and metaphorical space, and it is this false unity that we seek to open up and aerate, however cautiously. It is precisely the interconnectedness of metaphor and materiality that we seek to explore, not simply as a philosophical project but in order to advance in some way the shared project of a spatial politics.

METAPHOR AND FAMILIARITY

Metaphor works by invoking one meaning system to explain or clarify another. The first meaning system is apparently concrete, well understood, unproblematic, and evokes the familiar; in linguistic theory it is known as the 'source domain'. The second 'target domain' is elusive, opaque, seemingly unfathomable, without meaning donated from the source domain.[6] As Trevor Barnes and James Duncan have noted, drawing on the literary critic Kenneth Burke, metaphors assist in reducing the unfamiliar to the familiar;[7] they reinscribe the unfamiliar event, experience or social relation as utterly known. It is precisely this apparent familiarity of space, the givenness of space, its fixity and inertness, that make a spatial grammar so fertile for metaphoric appropriation.

Let us illustrate this simply by reference to three sets of concepts which are much in use in contemporary social and cultural discourse:

1 location, position and locality;
2 mapping;
3 colonization/decolonization.

In geographical terms, 'location' fixes a point in space, usually by reference to some abstract co-ordinate system such as latitude and longitude. Location may be no more than a zero-dimensional space, a point on a map. 'Position', by contrast, implies location *vis-à-vis* other locations and incorporates a sense of perspective on other places, and is therefore at least one-dimensional. 'Locality' suggests a two- (or more) dimensional place, an area within which multiple and diverse social and natural events and processes take place.[8] The appeal of these concepts as source domains for metaphors obviously lies in the precision and fixity they impute to the target domain. Notions like subject position, social location and locality borrow this concreteness of spatial definition to impose some order on the seemingly chaotic *mélange* of social difference and social relations. 'Social location' gives differentiated social subjects a place to stand, rendering them at the very least visible in their differences. 'Subject position' takes up the question of standpoint and the relativity of social location as a place of seeing and acting: different social actors, by virtue of their distinctive identities, are particularly located *vis-à-vis* other actors, and therefore enjoy a distinctive perspective from which they construct different social meanings.[9] The metaphor of locality, finally, suggests that social location is less an individual than a multi-dimensional experience, a collective engagement of mutually implicated identities.

Mapping provides a second illustration. In cartography, as Andrews says, 'space is used to represent space' (Andews 1990: 16). Mapping assumes a particular space as given; the function of mapping is to produce a scale representation of this space, a one-to-one correspondence between representation and represented, such that the outcome – the representation – is considered 'accurate' for some specified purpose. 'Mapping' is therefore a particularly useful metaphor for defining an area of enquiry or giving new form to a particular problem or set of problems. And, indeed, these are boom years for cartographic metaphors; theories, projects, concepts and differences are all being mapped.

Unlike location, position and locality – which all refer to specific spaces – mapping is an active process whereby the locations, structures and internal relations of one space are deployed in another. Typically, the replete material space of a given locality is represented on a cartographic space that is converted from a blank sheet of paper to a finished map. There are many ways to map a given space – none automatic, all requiring a substantive translation from the mapped to the map – and the value of such representations is traditionally measured in terms of its correspondence with a naïvely assumed 'reality'. In so far as mapping involves exploration, selection, definition, generalization and translation of data, it assumes a range of social cum representational powers, and as the military histories of geography and cartography suggest, the power to map can be closely entwined with the power of conquest and social control.[10] Although geographers and cartographers habitually give lip service to the selectiveness involved in mapping and to the realization that maps are strategic social constructions, they more often proceed in practice from traditional realist assumptions. Only recently have a few geographers and cartographers begun a sustained, theoretically informed and socially rooted (as opposed to simply technical) critique of cartographic conventions of positioning, framing, scale, absence and presence on the map, and, a critique of the absent if omniscient cartographer. Today the class, race, gender and national dimensions of mapping, broadly conceived, represent a vibrant field of enquiry,[11] and these same insights can illuminate the refractions of metaphorical mappings.

Third and finally, metaphors of colonization and decolonization are deployed to convey the dynamics of social domination in the everyday lives, thoughts and practices of social groups and historical subjects. In historico-geographical terms, colonization involves the conquest, inhabitation, possession and control of a territory by an external power. It is predicated on the deliberate, physical, cultural and symbolic appropriation of space. The violent and plunderous colonization of the non-European world over the last five centuries facilitated the ascendence of 'Western' hegemony and the global reach of capitalism. Decolonization following the Second World War marked the withdrawal of immediate military and political control by the European powers but, it is important to add, it did not result in the dilution of political-economic domination exercised through the world market and attendant institutions. Metaphors of colonization rescript this territorial incursion as an invasion and insidious habitation of the social and psychic space of oppressed groups, while decolonization becomes a metaphor for the process of recognizing and dislodging dominant ideas, assumptions and ideologies as externally imposed – literally of making a cultural and psychic place of one's own.

The obviousness and simplicitly of these translations from spatial to social identity and difference contrasts starkly with the purchase of spatial metaphors themselves. This is no accident but precisely the point. The more obvious the connection between source and target domain, the more revealing the metaphor may be. But metaphor does not work in quite the unidirectional manner we have so far assumed. The unfamiliar is variously distilled into the familiar, but this can have the reciprocal effect of revealing the familiar as not necessarily so familiar. The consistent and thematic use of metaphor, then, may better resemble Alice's passage through the looking-glass and back: the strange is rendered familiar, but the apparently familiar is made equally strange.

This is the case with contemporary spatial metaphor. The spaces and spatial practices that serve current metaphors in social, cultural and political theory are neither so fixed nor so unproblematic as their employment as metaphor would suggest. But before exploring the implications of this spatialization of language, we shall first attempt to sketch some of the sources of spatial metaphor in contemporary social theory.

SOURCES OF METAPHOR

In so far as metaphors are inherent to language and conceptualization – and spatial metaphors are no exception – the sources of spatial metaphor are at once everywhere and nowhere. But the use of spatial metaphors in contemporary social, political and especially cultural and literary discourse emanates from a more specific set of sources and circumstances, and here we select two such sources that have been central in very different ways: first, French structuralist and post-structuralist theory as exemplified by Althusser and Foucault; second, the comparative backwardness of spatial discourse in this century, especially in the English-speaking world.

Althusser and Foucault

Louis Althusser's radically structuralist rewriting of Marx represents an early and ambitious explication of social theory as spatial metaphor. History, for Althusser, was a 'process without a subject', and into the vacuum of a missing subject he inserted an array of performing spatial metaphors. The dialectic he modelled as spatial: presence/absence becomes a question of fields of vision; the visible/invisible dialectic is rendered as inside/outside a space; and spaces are treated as empty, homogeneous fields amenable to occupation by natural, social or conceptual events. As such, any space is 'limited by *another space outside it*'.[12] 'Theoretical space' is Althusser's way of evoking the possibility of new ideas, diametrically removed from ideological space. The 'metaphors in which Marx thinks', he argues, 'suggest the image of a change of terrain and a corresponding change of horizon'.[13]

Althusser's larger vision is equally spatial and metaphorical. His structural epistemology, connecting the raw material of knowledge, theoretical production, and the products of knowledge, he renders diagrammatically as a structure of connected 'levels'; and elsewhere, he justifies a modified base/superstructure distinction in terms of the architecture of a building. In a particularly Kantian vision of the intellectual division of labour, he identifies three 'great scientific continents' of knowledge: 'The continent of Mathematics . . . The continent of Physics' and 'the continent of History'. The latter represents the new terrain opened up by Marx. Internally, each of these continents of knowledge is divided into disciplinary 'regions'. Far from exploring a class politics of knowledge, Althusser envisages intellectual critique as an exploration and colonization process guided by the ad hoc logic of a kind of intellectual geopolitics.

Althusser reflected explicitly on his use of spatial metaphors to convey the social structure of capitalism. He argued that such metaphors are descriptive more than theoretical but a vital 'first phase of every theory'. Further, he identified what he saw as a central contradiction in his own use of spatial metaphor:

> The recourse made in this text to spatial metaphors (field, terrain, space, site, situation, position, etc.) poses a theoretical problem: the problem of the validity of its *claim* to existence in a discourse with scientific pretensions. The problem may be formulated as follows: why does a certain form of scientific discourse necessarily need the use of metaphors borrowed from non-scientific disciplines.
>
> (Althusser 1970: 26)

Althusser's structuralism has largely passed into history, but the spatial metaphors through which it was delivered live on in social and cultural theory. Perhaps because of his pretensions to scientific status, or perhaps because they did not alter his own 'theoretical practice', Althusser's well-founded if occasional misgivings about spatial metaphor have not been picked up. The metaphors have been carried forward without critical qualification.

Michel Foucault has also been an influential source of spatial metaphors; indeed, by his own account, spatial metaphors were an 'obsession' for him. He came to feel that more than any other grammar, spatial metaphors allowed him to express:

> the relations that are possible between power and knowledge. Once knowledge can be analysed in terms of region, domain, implantation, displacement, transposition, one is able to capture the process by which knowledge functions as a form of power and disseminates the effects of power.
>
> (Foucault 1980: 69)

Foucault persistently explored the connections between knowledge, power and spatiality, and maintained that the transition from temporal to spatial metaphors enabled a discursive shift from the realm of individual consciousness to wider 'relations of power' as constitutive of social meaning. For Foucault, the exercise of social power through the state and other social institutions and the exercise of power inherent in social opposition – this web of social power is modelled as a spatial field described by spatial strategies and geo-strategic interests. Like Althusser, Foucault was well aware of his reliance on spatial metaphors, and at least once he reflected critically on the devaluation of space in intellectual discourse, the supposed deadness, fixity and immobility of space, while emphasizing its fecundity for expressing relations of power.[15]

Foucault seems to have felt that his choice of spatial metaphors served to redress the mistaken privilege of time *vis-à-vis* space, and encouraged the task of 'making the space in question concrete', yet he seems not to have grasped the full power of his own use of metaphors. In the first place, Foucault's pervasive substitution of spatial metaphor for social structure, institution and situation continues to elide the agency through which social space and social relations are produced, fixing these instead as the outcome of juridico-political forces. And far from enlivening a spatial discourse, Foucault invariably occludes the actual spatial source of such metaphors as domain, field, region, even territory and imperialism, thereby excising the possibility of examining the mutual translations between geographical and metaphorical space. In Foucauldian terms, the 'space' between material and metaphorical space is radically denied, and the collapse of this distinction is marked by a deeply ambivalent dismissal cum retention of geographical space. Thus he denies that an investigation of space and spatial matters – the source of his metaphors – is of any interest and relegates such an enquiry to a disciplinary project of 'the archeology of geography'; and yet at the same time, in suggesting that 'Geography must indeed necessarily lie at the heart of [his] concerns', Foucault eagerly anticipates ransacking political geography in search of fresh metaphors:

> One theme I would like to study in the next few years is that of the army as a matrix of organisation and knowledge; one would need to study the history of the fortress, the 'campaign', the 'movement', the colony, the territory.
>
> (Foucault 1980: 77)

As the interviewer from *Herodote* phrased it in this discussion with Foucault, 'the recourse to spatial metaphors' may be necessary, but it is 'regressive, [and] non-rigorous' (ibid.: 70). Foucault, who so brilliantly excavated the deployment of power in the institutions of everyday life and the mundane practices associated with them, fails to recognize how social agents produce space and socio-spatial relations within and against the economic, political and juridical imposition of produced space and spaces. If, as Spivak suggests, Foucault's 'concern for the politics of the oppressed' disguises a 'privileging of the intellectual' – indeed the Western intellectual – then his use of spatial metaphors contributes to this result:

> Foucault is a brilliant thinker of power-in-spacing, but the awareness of the topographical reinscription of imperialism does not inform his presuppositions. . . . The clinic, the asylum, the prison, the university [for Foucault] all seem to be screen-allegories that foreclose a reading of the broader narratives of imperialism.
>
> (Spivak 1988: 292)

Underdevelopment of spatial discourse

The use of spatial metaphors is, ironically, encouraged by the very fact that spatial discourse has been so underdeveloped. Precisely in its deadness, as Foucault puts it – the taken-for-grantedness of spatial meanings – the language of space becomes a fertile repository or source domain for metaphor. In the short term, the ossification of space and spatial concepts has evolved over much of the twentieth century, as socio-spatial concepts in the social sciences (geography notwithstanding) have failed to develop along with concepts of space in physics or art. If the discipline of geography has been in the forefront of the recent reassertion of space, it was also a central contributor to the death of space. The relativity and relationality of space explored in physics and art were of little concern to geographers throughout most of this century. Rather than contribute to these scientific and cultural developments, geographers clung to a traditional (Newtonian) conception of absolute space – space as field or container, describable by a two- or three-dimensional metric of co-ordinates. From the First World War to the 1960s, geographical research in Europe and North America emphasized descriptive studies of social and natural forms located in absolute space and the connections between landscape forms and natural processes. Social process was rarely considered in any serious way, but when it was – as, for example, in economic or political geography – social processes were conceived as connecting pre-existing spaces or as happening 'in' or 'across' an equally given spatial field. Geographers recoiled from engaging emerging social theory from Weber to Durkheim to the Frankfurt School, not to mention work derivative of Marx or Freud, and by mid-century they had become increasingly isolated from other social sciences. As Tom Glick once said of mid-century American geography, undoubtedly the most isolationist, 'it is middle American, middle class and middle brow'.[16]

As geographers' treatment of space stagnated, social theory and the social sciences more broadly refused responsibility or interest in comprehending or exploring questions of social space, and despite the invention of new languages of space in mathematics, physics and art, the language of social space ossified. Precisely in their refusal to explore alternative conceptions of social and physical space, geographers contributed to the deadening of space that prepared the way for, or at least accentuated, the power of spatial metaphor.

Absolute space

Spatial metaphors are problematic in so far as they presume that space is not. And they are problematic in political as much as philosophical terms. The problem lies not with spatial metaphors as such, but with metaphors that depend on a very specific representation of space: *absolute space*. 'Absolute space' refers to a conception of space as a field, container, a co-ordinate system of discrete and mutually exclusive locations. Absolute space is the space that is broadly taken for granted in Western societies – our naïvely assumed sense of space as emptiness – but it is only one of many ways in which space can be conceptualized. The representation of space as absolute also has a very specific history. Prefigured in Euclid's geometry, of course, absolute space was widely established as a dominant representation of space between the seventeenth and nineteenth centuries. In the first place, an absolutist conception began to dominate the philosophical and scientific discussion of space (at precisely the time that Greek thought was being rediscovered as the apparent progenitor of 'Western thought') with the work of Newton, Descartes and Kant. Space was infinite and a priori; it was geometrically divisible into discrete bits; and, at least in Newton's case, it was the empirical proof of an omnipresent god. Though certainly contested by thinkers such as Leibnitz, who proposed to see space in relational terms, absolute space became increasingly hegemonic.[17]

Equally as important and closely connected, absolute space did not become the lingua franca of science in a social vacuum. The emergence of capitalist social relations in Europe brought a very specific set of social and political shifts that established absolute space as the premise of hegemonic social practices. The inauguration of private property as the general basis of the social economy, and the division of the land into privately held and precisely demarcated plots; the juridical assumption of the individual body as the basic social unit; the progressive outward expansion of European hegemony through the conquest, colonization and defence of new territories; the division of global space into mutually exclusive nation-states on the basis of some presumed internal homogeneity of culture (albeit a division brought about with economic motivation and through military force): these and other shifts marked the emerging space-economy of capitalism from the sixteenth century onwards and represented a powerful enactment of absolute space as the geographical basis for social intercourse.

More than any other, the absolute conception of space has contributed to what is widely assumed today in Western societies as 'real space', the space of contemporary 'commonsense'. And despite its apparently abstract neutrality, absolute space is politically charged in its contemporary implications as much as in its historical origins. As Lefebvre emphasizes, it is a conception of space appropriate for a project of social domination.[18] Foucault is correct to see the connection between space and power, therefore, but indulges in a reductive universalization of this connection – a universalization, incidentally, that allows his spatial metaphors to take on something of a free floating existence that denies their referents and material results. It is not space *per se* that expresses power, but the thoroughly naturalized absolute conception of space that grew up with capitalism, and which expresses a very specific tyranny of power. The power of Einstein's relativity theory, the power of cubism and surrealism was not simply that they overcame established scientific or artistic conceptions of space but that they fundamentally challenged the absolutist conception on which a wider web of social, economic, military and cultural relationships were modelled.[19]

THE POWER OF SPATIAL METAPHOR

Adrienne Rich's 'Notes toward a politics of location' was a highly influential text which coalesced a range of feminist debates around the question of how 'we' can see the world politically from our multiple social perspectives, and who, in any case, 'we' are. It focused much of the discussion concerning how to conceptualize gendered social locations. She suggests from the start the need for a spatially scaled sense of social identity and begins where much feminist discussion has focused, with the political geography of the female body:

> And Marxist feminists were often pioneers in this work. But for many women I knew, the need to begin with the female body – our own – was understood not as applying a Marxist principle *to* women, but as locating the grounds from which to speak with authority *as* women. Not to transcend this body but to reclaim it.
>
> (Rich 1986: 213)

The metaphor of location here serves very elegantly in the accomplishment two interrelated political shifts. It quite literally renders her, Adrienne Rich, visible; it gives her a place to stand, a specific place. But an inclusionary rather than exclusionary metaphor, it does the same for women as a group; it makes women visible as differently constituted social actors – older, Latina, Canadian, lesbian – and claims in the name of women a place of our own. If it did only this, the locational metaphor would have been fruitful but not necessarily lasting or especially revealing, and its authority would have remained firmly dependent on the absoluteness of spatial location. But Rich goes further and accomplishes a second shift. By returning to the scale of her self, she challenges the presumed homogeneity of identities: 'the meaning of my whiteness', she says, is 'a point of location for which I need to take responsibility'. Rich recognizes the relativity of location in geographical cum social terms, and much of the rest of the essay represents an exploration of this theme, citing at one point 'a Third World poster' which reads: 'WE ARE HERE BECAUSE YOU WERE THERE.'[20]

Rich's success at disrupting the imposed social mapping of identities is dependent not simply on claiming a location for women, for black women, for Jewish lesbians, and so forth, but on questioning the very process through which the base map of different locations is drawn and through recognizing that the relationality of *social* location is inextricably imbricated with the relationality of geographical location. Further, the relativity of location applies not just to us versus them, one group *vis-à-vis* another, but inevitably implies a redefinition of ourselves, of the group. Just as she maintains that 'the United States has never been a white country', she concludes by asking who the 'we' are who are the supposed subject of political change.[21]

The brilliance of Rich's formulation is that it maintains the relationality of social identity without slipping into a formless relativism, and at the same time disarranges the received fixity of social *and* geographical location. Unfortunately, this double shift is not always repeated among those who have drawn inspiration from Rich. The richness of metaphorical insight is reduced to a conceptual one-dimensionality when the first shift she makes is not accompanied by the second. Among the cruder versions of standpoint theory, for example, or in the simplest statements of 'perspectival difference', the absoluteness of geographical space is never questioned, and the political ramifications filter back to infect the map of the social: social location, inherently fluid, is inadvertently mapped as absolute. Society is implicitly rendered as a *mappa mundi*, a blank space on to which social locations are projected – a New World of sorts, ready for colonization; identities are located, positioned, elbowed into an already existing social mosaic. Identity politics too often *becomes* mosaic politics; at its worst it quickly devolves towards an unseemly nationalism of competing isms – a geopolitics of identity which is still being fought out.

The conceptual and political revolt, the liberation of located, scaled and interwoven identities that Rich and others seek is thereby frustrated in part by the unseen implications not of spatial metaphor as such but by metaphor whose meaning is donated by an absolutism of space; by an unwillingness to rethink simultaneously the spatial substance of such metaphors or an unawareness of their operation. The taken-for-grantedness of space is indeed a political issue.

The apparent cul-de-sac of a strict identity politics is now quite evident, and has led directly to a softening of the spatial metaphors through which political location is mapped. The discussion of multiple identities, borders and borderlands, margins, and escape from place are all in different ways a response to the political inviability of absolute location. For the most part, however, these alternative metaphors do not go far enough. 'Multiple identities' need go little further than implying a map overlay of social experience in which the composite result is or is not greater than the sum of its parts. The notion of margins and borderlands is more interesting, especially with the implication of permanent location at the edge, but of course it leaves a core identity intact, a forceful locus of power uninterpolated. Absolute space is perhaps marginalized but its critical *Aufhebung* is barely initiated.

The notions of travel, travelling identities and displacement represent another response to the undue fixity of social identity. 'Travelling' provides a means for conceptualizing the interplay among people that are no longer so separate or inaccessible one to the other. Travel erodes the brittleness and rigidity of spatial boundaries and suggests social, political and cultural identity as an amalgam, the intricacy of which defies the comparative simplicity of 'identity'. The flow of travel not the putative fixity of space donates identity. Concerned primarily with travelling theory and travelling cultures, James Clifford has been among the most thoughtful proponents of travel metaphors. A 'term of cultural comparison' and a means of 'comparative knowledge', travel for Clifford moves us beyond the fixity of singular locations, as he suggests in explicit relation to Adrienne Rich:

> 'Location,' here, is not a matter of finding a stable 'home' or of discovering a common experience. Rather it is a matter of being aware of the difference that makes a difference in concrete situations, of recognizing the various inscriptions, 'places' or 'histories' that both empower and inhibit the construction of theoretical categories like 'Woman,' 'Patriarchy,' or 'colonization,' categories essential to political action as well as to serious comparative knowledge. 'Location' is thus, concretely, a *series* of locations and encounters, travel within diverse, but limited spaces. Location, for Adrienne Rich, is a dynamic awareness of discrepant attachments – as a woman, a white middle-class writer, a lesbian, a Jew.[22]

With the notion of travel, Clifford reaches for a dynamic rather than static conception of location. He wants to convince anthropologists and others that the traditional disciplinary practice of 'localizing' cultures in 'the field' or 'the village' is problematic in that identities are established in the course of travel as much as spatial and cultural rootedness. In so doing, he reinstates a usable cosmopolitanism, as Bruce Robbins puts it (Robbins 1992: 181), if one tempered by the resilience of local specificity. But as his discussion here and elsewhere suggests, Clifford comprehends and accomplishes only the first shift of Rich's 'politics of location', remaining effectively blind to the second. It is not simply 'a dynamic awareness of discrepant attachments' that Rich suggests – a vague misformulation which admits multiple locations but leaves existing locations rigidly in place and only abstractly connected; there is no glimpse here of the combined rupture of received social/geographical space. Clifford's '*series* of locations' and 'diverse, but limited spaces' suggest a multiplication of absolute spaces rather than a radical rethinking of spatial concepts. Travel for Clifford 'takes place' over a pre-given space or series of spaces, a more replete map of social location than the *mappa mundi* to be sure, an already partly filled in space, but none the less pre-given. Travelling theorists and cultures are very much changed in the course of their travel, they change other travellers, and they change the people they meet, but the *places* they travel to and from remain unaltered. The point here is that the 'natives' of anthropological enquiry cannot be so readily separated from their place.

'The metaphor of travel, for me', Clifford says, 'has been a serious dream of mapping without going "off earth" ' (Clifford 1992: 105) – a kind of comparative world cultural geography presumably. While this notion of travel succinctly expresses the relationality inherent in the web of social life, the complex connectedness of identity and identities, this is achieved, paradoxically, via a radical reassertion of the deadness of space, the fixity of an underlying spatial structure, that may none the less take us 'off earth'.[23] The subject moves but space stands still, fixed, unproduced.

Indeed, it is tempting to think of Clifford's travellers as the frequent fliers of theory and knowledge, accumulating culture along with mileage while gazing down at the patterned land below. Diverse as space and spaces are, 'space' itself is rendered unproblematic, in startling contrast to the 'everything flows' of the social. The hard if largely hidden reassertion of an absolutist spatial ontology performs as virtual image for the social flux; spatial language comes to ground social meaning. It provides the missing foundation for everything else in flux. Without the pre-given structure of diverse if limited spatial locations, to which and from which our travellers travel, there is nothing to stop the flux of 'travel' from deteriorating into complete relativism. This Clifford does not do, of course; as he himself says, this 'is not nomadology' (Clifford 1992: 108). Yet the price paid for his limited dynamicization of socio-spatial relations may be too high. Certainly, Adrienne Rich's ambition for a disruptive politics of location is unhinged. To escape the potential essentialism of overly fixed social identities, he imports a spatial essentialism connected with the most modernist of spatial concepts. The unexamined silences of spatial metaphors, then, may covertly constrain the formation of the very political alliances and possibilities they seem to invoke.

CONCLUSION

Newton, Descartes and Kant were the philosophical progenitors of spatial modernism, as much as Columbus, Napoleon and the Duchess of Sutherland were its practitioners. The depth of their collective influence, the taken-for-grantedness of the absolute space they established, is only beginning to be challenged. That this space is quite literally the space of capitalist patriarchy and racist imperialism should hasten critique and reconstruction. Meanwhile the uncritical appropriation of absolute space as a source domain for metaphors forecloses recognition of the multiple qualities, types, properties and attributes of social space, its constructed absolutism and its relationality. This is not to say, therefore, that absolute space has no real referent; in modern representations of the body, private property, the state and colonization, absolute space is very real if socially constructed. The problem lies rather in the naturalization of absolute space which leads, in turn, to a tendency for such metaphors to become virtually free-floating abstractions, the source of their grounding unacknowledged. The widespread appeal to spatial metaphors, in fact, appears to result from a radical questioning of all else, a decentring and destabilization of previously fixed realities and assumptions; space is largely exempted from such sceptical scrutiny precisely so it can be held constant to provide some semblance of order for an otherwise floating world of ideas. Virtually universal in social discourse, absolute space has become unhinged from the actual spatial experiences that rendered it an appropriate social conception. To the extent that this occurs, we are seeing, in the words of Dick Hebdige, 'metaphors out of control' (Hebdige, forthcoming).

Henri Lefebvre has made a parallel critique. In the early 1970s, discussing various social theorists (including Foucault and Althusser) whose influence on social and cultural theory was expanding outward from France, he argued:

> Most if not all of these authors ensconce themselves comfortably enough within the terms of mental (and therefore neo-Kantian or neo-Cartesian) space, thereby demonstrating that 'theoretical space' is already nothing more than the egocentric thinking of specialized Western intellectuals – and indeed may soon be nothing more than an entirely separated, schizoid consciousness.[24]
>
> (Lefebvre 1991: 24)

Lefebvre was of course an *aficionado* of metaphor himself, but not of free-floating metaphor that denies its own construction. When space itself is seen as amenable to social construction, the taken-for-grantedness of the source domain for spatial metaphors is denied, and the metaphors are opened to scrutiny. The point here is to enable an explicit consideration of the meaning and politics of metaphor by investigating the connectedness, the imbrication of material and metaphorical space. Our argument is not against spatial metaphors as such; rather that by recognizing that space *is* problematic, we seek to undermine some of the assumptions that render spatial metaphors seemingly reflexive and thereby to enrich the reservoir of meanings available for politicized spatial metaphor.

NOTES

1 On the reassertion of space, see Soja (1989).
2 See Jameson (1984: 53–92), Harvey (1989); Ross (1988); Probyn (1990); Castells (1991); and Smith (1991).

3 Kern (1983); Schorske (1980).
4 See also Harvey (1973); Lefebvre (1991; 1976).
5 Jameson (1984; 1989).
6 The language of source domain and target domain comes from Naomi Quinn (1991).
7 Barnes and Duncan (1992: 11).
8 Even in the geographical literature, the definition and implications of localities are not always comprehended. See, for example, the debate on 'locality studies' carried out in the pages of *Antipode* and elsewhere. See, *inter alia*, Smith (1991: 59–68); Cox and Mair (1989: 121–32); Duncan and Savage (1989: 179–206).
9 See, for example, Probyn (1990); Harding (1986; 1990).
10 Harley (1989; 1990); Wood (1992); Pickles (1992).
11 In geography, see especially the work of Brian Harley (1989; 1990), whose tragic death in 1991 robbed us of a vital intellectual bridge between geography, cartography and cultural studies. See also Wood (1992) and Pratt (1992).
12 Althusser (1970: 26), emphasis in original; see also ibid.: 182.
13 Althusser (1970: 24).
14 Althusser (1971: 138); *Politics and History*, p. 166.
15 Foucault (1980: 68–71); see also Foucault (1979; 1982).
16 Glick (1983). See also J. N. Entrikin and S. Brunn (1989).
17 See Max Jammer (1969).
18 Lefebvre (1991).
19 See Smith (1991, Ch. 3).
20 Rich (1986: 219, 226).
21 Rich (1986: 226, 231). See also Anzaldua (1990) and Marston (1990).
22 Clifford (1989: 82; 1992: 110); Said (1983).
23 See especially the discussion of 'deep space' in Smith (1991: 160–5).
24 Cf. Adrienne Rich: 'Even in the struggle against free-floating abstraction, we have abstracted. Marxists and radical feminists have both done this. Why not admit it, get it said, so we can get on to the work to be done back down to earth again?' (Rich 1986: 218–19).

REFERENCES

Althusser, L. (1971) *Lenin and Philosophy and Other Essays*, London, New Left Books.
—— (1975) *Reading Capital* (1st edn 1970), London, New Left Books.
Andrews, J. H. (1990) 'Map and language: a metaphor extended', *Cartographica* 27.
Anzaldua, G. (1990) 'La conciencia de la Mestiza: towards a new consciousness', in G. Anzaldua (ed.) *Making Face, Making Soul, Haciendo Caras. Creative and Critical Perspectives by Women of Color*, San Francisco: Aunt Lute Foundation Books, 377–89.
Barnes, T. J. and Duncan, J. S. (1992) 'Introduction', in T. J. Barnes and J. S. Duncan (eds) *Writing Worlds: Discourse, Text and Metaphor in the Representation of Landscape*, London: Routledge, 1–17.
Brunn, S. and Entrikin, J. N. (eds) (1989) *Reflections on Richard Hartshorne's 'The Nature of Geography'*, Washington, DC: occasional publication of the Association of American Geographers.
Castells, M. (1991) *The Informational City*, Oxford: Basil Blackwell.
Clifford, J. (1989) 'Notes on travel and theory', in J. Clifford and V. Dhareshwar (eds) *Inscriptions 5: Traveling Theories, Traveling Theorists*, Santa Cruz Center for Critical Studies, Santa Cruz: University of California.

—— (1992) 'Travelling cultures', in L. Grossberg, C. Nelson and P. Treichler (eds) *Cultural Studies*, New York: Routledge, 96–112.
Cox, K. and Mair, A. (1989) 'Levels of abstraction in locality studies', *Antipode* 21: 121–32.
Duncan, S. and Savage, M. (1989) 'Space, scale and locality', *Antipode* 21: 179–206.
Foucault, M. (1979) *Discipline and Punish: the Birth of the Prison* New York: Vintage.
—— (1982) 'Space, knowledge and power', in *The Foucault Reader*, ed. P. Rabinow, New York: Pantheon, 239–56.
—— (1980) 'Questions on Geography', in *Power/Knowledge: Selected Interviews and Other Writings 1972–1977*, ed. C. Gordon, New York: Pantheon, 63–77.
Glick, T. (1983) 'In search of geography', *ISIS* 74(271): 92–7.
Harding, S. (1986) *The Science Question in Feminism*, Ithaca, NY: Cornell University Press.
—— (1990) 'Feminism, science and the anti-Enlightenment critiques', in L. J. Nicholson (ed.) *Feminism/Postmodernism*, London: Routledge, 83–106.
Harley, J. B. (1989) 'Deconstructing the map', *Cartographica* 26(2): 1–20.
—— (1990) 'Cartography, ethics and social theory', *Cartographica* 27(1): 1–23.
Harvey, D. (1973) *Social Justice and the City*, London: Edward Arnold.
—— (1989) *The Condition of Postmodernity*, Oxford: Basil Blackwell.
Hebdige, D. (forthcoming) 'Metaphors out of control', in J. Bird *et al.* (eds) *Mapping the Futures*, London: Routledge.
Jameson, F. (1984) 'Postmodernism, or, the cultural logic of late capitalism', *New Left Review* 146: 53–92.
—— (1989) 'Marxism and Postmodernism', *New Left Review* 176: 31–45.
Jammer, M. (1969) *The Concept of Space*, Cambridge, Mass.: Harvard University Press.
Kern, S. (1983) *The Culture of Time and Space, 1880–1918*, London: Wiedenfeld & Nicolson.
Lefebvre, H. (1976) 'Reflections on the politics of space', *Antipode* 8; reprinted in R. Peet (1978) *Radical Geography: Alternative Viewpoints on Contemporary Social Issues*, London: Methuen, 339–52.
—— (1991) *The Production of Space*, Oxford, Basil Blackwell.
Marston, S. A. (1990) 'Who are "the People"?: gender, citizenship, and the making of the American nation', *Environment and Plannning D: Society and Space* 8(4): 449–58.
Pickles, J. (1992) 'Texts, hermeneutics and propaganda maps', in T. J. Barnes and J. S. Duncan (eds) *Writing Worlds: Discourse, Text and Metaphor in the Representation of Landscape*, London: Routledge, 193–230.
Pratt, M. L. (1992) *Imperial Eyes: Travel Writing and Transculturation*, London: Routledge.
Probyn, E. (1990) 'Travels in the Postmodern: making sense of the local', in L. J. Nicholson (ed.) *Feminism/Postmodernism*, New York: Routledge, 176–89.
Quinn, N. (1991) 'The cultural basis of metaphor', in J. W. Fernandez (ed.) *Beyond Metaphor: the Theory of Tropes in Anthropology*, Stanford, CA: Stanford University Press.
Rich, A. (1986) 'Notes toward a politics of location', in A. Rich, *Blood Bread and Poetry: Selected Prose, 1979–1985*, New York: W. W. Norton.
Robbins, B. (1992) 'Comparative cosmopolitanism', *Social Text* 31/32.
Ross, K. (1988) *The Emergence of Social Space: Rimbaud and the Paris Commune*, Minneapolis: University of Minnesota Press.
Said, E. (1983) 'Traveling Theory', in E. Said, *The World, the Text, and the Critic*, Cambridge, Mass.: Harvard University Press, 226–47.
Schorske, C. (1980) *Fin-de-siècle Vienna: Politics and Culture*, New York.
Smith, N. (1987) 'Dangers of the empirical turn: some comments on the CURS initiative', *Antipode* 19: 394–406.

—— (1991) *Uneven Development* (2nd edn), Oxford: Basil Blackwell.
Soja, E. (1989) *Postmodern Geographies: the Reassertion of Space in Critical Social Theory*, London: Verso.
Spivak, G. C. (1988) 'Can the subaltern speak?', in L. Grossberg and C. Nelson (eds) *Marxism and the Interpretation of Culture*, Urbana: University of Illinois Press, 271–313.
Wood, D. (1992) *The Power of Maps*, New York: Guildford Press.

5

LOCATING IDENTITY POLITICS

Liz Bondi

Identity politics has had a mixed press. In support Rosalind Brunt describes it as 'politics whose starting point is about recognising the degree to which political activity and effort involves a continuous process of making and remaking ourselves – and ourselves in relation to others' (Brunt 1989: 151). Disparagingly, Jenny Bourne complains that this emphasis entails a retreat from collective, emancipatory, political projects: 'Identity Politics is all the rage. Exploitation is out (it is extrinsically determinist). Oppression is in (it is intrinsically personal). What is to be done is replaced by who am I' (Bourne 1987: 1).[1]

In this chapter I examine what kind of entity 'identity politics' might be. In the context of these different appraisals, I shall consider where identity politics is to be located on a map of intellectual traditions, and what this kind of geographical metaphor contributes to identity politics. My discussion is prompted in part by the history of my own involvement in feminist politics in Britain,[2] but I believe the issues that I address are relevant to radical politics in general.

In the first part of this chapter, I elaborate on the contrasting interpretations offered by Brunt and by Bourne, associating the first with an anti-humanist impulse I trace to Marx and Freud, and associating the second with an apparent reassertion of liberal humanism within oppositional political forms. In the remainder of the chapter, I argue that, in practice, a tension between anti-humanism and humanism, between the idea of the human subject as a social construct or as rooted in a pre-given essence, necessarily pervades identity politics. I identify Marx and Freud as anti-humanist sources for an emancipatory politics of identity in part as a reminder that the critique of liberal humanism began long before the emergence of post-structuralism.

But neither legacy directly generated such a politics, and in the second section I draw out some aspects of their respective formulations that militated against the development of a politics of identity. In both cases I exemplify relevant points in relation to questions of gender identity.

I then turn my attention to feminist politics where issues of identity have been particularly significant and contentious, and which, I believe, demonstrate the variegated character of identity politics with particular clarity. I point to some of the pitfalls of identity politics, manifest especially in notions of authentic identities and hierarchies of oppression. But at the same time, I acknowledge the need to resist claims that identity is infinitely fluid and malleable. In this context, the tension between humanism and anti-humanism has increasingly been expressed in a debate about essentialism, and, using examples from lesbian and gay politics as well as feminism, I emphasize the strategic importance of certain versions of essentialism. Finally, I examine the functioning of terms such as 'location' and 'positionality' in discussions of these issues and suggest that place may be the latest repository for essences in contemporary versions of identity politics.

IDENTITY POLITICS: PROGRESSIVE CHALLENGE OR REACTIONARY CELEBRATION?

In very different ways, both Marx and Freud set in motion a radical reconceptualization of the human subject.[3] Against notions of individual rationality, self-knowledge and self-mastery, both insisted on the overwhelming power of processes operating beyond the conscious control of individual agents, and on the inevitable fragility of taken-for-granted notions of self-hood. For Marx, the structure of class relations was paramount. His analysis of the alienation of workers from the fruits of their labour entailed a diagnosis of consciousness as a product, rather than a cause of human action. And he insisted that the individual of social contract theory and classical economics was a mythical construct that systematically misrecognized the nature of social relations. He implied that no one, bourgeois or proletarian, is truly in command of himself or herself, and that we occupy subject positions created by the conditions of class relations. Joel Kovel points to the parallel with Freud when he writes that, for Marx, 'consciousness, or immediate introspection, is a profoundly unreliable guide. Put more decisively, the Marxist believes in an unconscious, but not one within the mind' (Kovel 1988: 176–7).

For Freud, of course, psychical processes were paramount. He insisted that consciousness is constructed in response to the unconscious, which itself is the product of repressed wishes and desires, especially those of infancy. The unconscious is never directly knowable, but its insistent presence renders consciousness always in some ways self-deceiving, never wholly rational, and never wholly secure. From this perspective, the individual subject is not, and cannot be, a coherent, unified being, but is always divided and displaced. Against this, we struggle to sustain a myth of ourselves as self-determined agents, as the authors of our actions, which we can never wholly become. We exist in a state of psychic alienation. Again, Freud's analysis runs parallel to (and not convergent with) Marx's: 'Freud's work remorselessly undermined the pretensions of reason and common sense – indeed, showed them to exist in precisely the same kind of emergent relationship to the repressed past that Marx had shown for historical activity in general' (Kovel 1988: 173).

Each formulation offered a direct challenge to the assumptions and certainties of liberal humanism,[4] most obviously to the Cartesian *cogito* and its autonomous, sovereign subject, and also to the more general notion of an irreducible, stable, unalienated essence at the core of, and giving coherence to, every human individual. The idea of a self-sufficient unity within the human individual creates the basis for people to identify with one another as equals. Consequently, the liberal humanist view of the human subject has often been drawn upon to argue for such emancipatory necessities as the equal value of all human beings, and to endorse calls for equal rights. It remains the dominant hegemonic view in Western liberal democracies. But the model has turned out to be as much a mechanism for subordination as a counter to it, because its supposed universality rests upon a suppression of difference. Moreover, the differences that are suppressed are differences from those who occupy positions of power and who have the authority to define knowledge.[5] Thus, to qualify for equality means becoming indistinguishable from the authors of this viewpoint: white, Western, bourgeois men. And the spurious claim to universality effectively creates excluded, minoritized groups.[6]

In challenging this conception of the human subject, Marx and Freud opened up possibilities for resisting the normative claims (rather than the egalitarian ambitions) of liberal humanism. Both insisted that identity – our sense of ourselves as individuals and as social beings – is constructed through structural processes rather than being innate or pre-given. In so doing, both also implied that there are no necessarily universal or unchanging attributes of human identity, but that differentiation and movement between identities is characteristic of modern societies.

Subordinated groups can make use of these insights to insist on the fraudulence of the apparent self-mastery and authority of the bourgeois individual, to resist their positioning as 'others' or 'minorities', and to construct alternative identities as part of a politics of resistance or opposition. I would argue, therefore, that both Marx and Freud provide theoretical bases for a politics in which a counter-hegemonic construction of the human subject is marshalled to challenge dominant groups and to organize the subordinated, a politics in which an oppositional redefinition and assertion of identity is central.

This is one interpretation of what identity politics is about. It positions identity politics very much as part of an emancipatory politics of opposition. But the term 'identity politics' is also often used in a derogatory manner to refer to an emphasis on feelings and experiences that appears to *reinstate* a liberal humanist conception of the subject within contemporary oppositional politics. According to this view, identity politics is about the (re)discovery of an already existing identity. It is founded on a conception of the human subject as centred, coherent and self-authored, albeit masked over and distorted by the effects of subordination. It differs from the classic liberal humanist model only in that it takes identity to be plural: there exist several distinct identities, each to be recovered and celebrated. Within this interpretation, pre-given, essential identities define and determine politics. Because differences between people appear as innate and unbreachable, according to its critics, identity politics militates against the development of broadly based oppositional political movements. Moreover, its critics argue that the emphasis on self-discovery has the effect of replacing politics with therapy. Its adherents are also accused of succumbing to dominant ideas about individuality and authenticity: identity becomes a compensation for, rather than a challenge to, relations of exploitation.

How can this negative appraisal coexist with the idea that identity politics is about resisting and challenging the fraudulent claims of dominant groups? I offer answers to this question – first in terms of tensions within the theoretical frameworks to which I have attributed an emancipatory politics of identity, and second in terms of the practices of second-wave feminism.

FREUD AND MARX: DECENTRED SUBJECTS OR POLITICAL AGENTS?

To describe Freud as a source for oppositional politics may seem a little odd. A great deal of psychoanalysis is anything but counter-hegemonic. Most obviously, the early twentieth-century appropriation of Freud's ideas in the United States led to the emergence of an approach that eclipsed the subversive elements of psychoanalysis and that advanced therapeutic techniques designed to encourage compliant adaptation to societal norms.[7] This repositioned psychoanalysis, in the guise of ego-psychology, firmly within the domain of liberal humanism.

Freud sought to distance himself from such (mis)applications, but his own writing is not without ambiguity. Nowhere is this more apparent than in his treatment of women and feminine sexuality. While his account denaturalizes the condition of women, it is also deeply androcentric: he offers some remarkably powerful insights into the workings of patriarchy but fails to draw the existing social relations of gender into question, thereby implying their inevitability. As Jeffrey Weeks argues:[8]

> the problem with Freud . . . is that he constantly oscillates between this radical insight [the difficulty of femininity] and his own normalising tendency. . . . Freud is saying simultaneously that gender and sexual identities are precarious, provisional and constantly undermined by the play of desires, and that they are necessary and essential, the guarantee of mental and social health.
>
> (Weeks 1985: 144, 148)

This generates the possibility of reading Freud as either disclosing or endorsing the position of women as subordinated 'other'. Feminists leaning towards the former take Freud's insistence that femininity is not biologically given but psychically constituted, and his emphasis on the precariousness of identity, as opening up possibilities for the creation of a feminist, rather than feminine, subject: 'Feminism's affinity with psychoanalysis rests above all, I would argue, with this recognition that there is a resistance to identity at the very heart of psychic life' (Rose 1986: 91). In contrast, those leaning towards the latter, view psychoanalysis as a deeply insidious ploy that intensifies women's subordination and defuses women's resistance to patriarchy, leading Elizabeth Wilson to argue that

> the last thing feminists need is a theory that teaches them only to marvel anew at the constant recreation of the subjective reality of subordination and which reasserts male domination more securely than ever within theoretical discourse.
>
> (Wilson 1987: 175)

Freud was equally ambiguous on the status of heterosexuality: on the one hand, he denaturalizes it and instates bisexuality as a natural condition of childhood; on the other hand, he normalizes adult heterosexuality and positions homosexual object choice as perverse. Psychoanalytic practice since Freud has overwhelmingly regarded gay and lesbian sexualities as pathological, curable conditions, so that within gay and lesbian politics, psychoanalysis is generally regarded with deep mistrust and scepticism (see Weeks 1985).

This slippage 'from analysis to prescription' (see Weeks 1985: 148) precluded the development of a politicized notion of subjectivity in which the precariousness of gender and sexual identities might be deployed to challenge societal norms. But this is not just a matter of the mode of assimilation of Freud's ideas: in at least two different and contradictory ways Freud himself is not merely ambivalent but actively resistant to a politically committed reading.

First, Freud retained a commitment to scientific objectivity, rationality and to a sharp distinction between fact and value, which positioned such a project entirely outside his intellectual endeavour. In practice, this form of resistance is relatively easily overcome. Freud's scientific aspirations conflict with the anti-humanism intrinsic to his theory of subjectivity: in drawing upon a notion of the scientist, he posits a rational, detached observer and in this way readmits the subject-in-command associated with liberal humanism. Thus, although theoretically committed to the idea of a subject that is constructed through unconscious processes and that cannot be consciously known or controlled, Freud did not unsettle his own position as authoritative speaker. But he did provide the means by which this could be done (see, for example, Flax 1990).

Second, and more importantly, through his insistence on the unconscious, Freud posited a subject 'radically *incapable* of knowing itself' (Grosz 1990: 13), and therefore he brought into question the possibility of consciously constructing oppositional identities sufficiently stable to sustain emancipatory political projects. Attempts to overcome this problem by uniting Freud with Marx have generally downplayed the role of the unconscious. However, this means relinquishing the dynamism of Freud's theory, especially the powerful and destabilizing role of desire, and offering overly deterministic theories.[9] Lacan's influential return to Freud reinstates the unconscious eclipsed by critical theory, ego-psychology and object relations theory alike. But his reading of Freud insists on the unremittingly alienated and self-deceptive qualities of consciousness in a manner that seems to imply that emancipatory goals are inherently misguided. He was especially disparaging about the aspirations of feminism.

In contrast to Freud, for Marx, the whole point of a critique of pre-given consciousness was to advance the possibility of an oppositional political subject. Like Freud, his account (explicit or implicit) of the human subject is not singular or unitary but is open to different readings. Here I draw on Marshall Berman's reading of the *Communist Manifesto*, which pays particular attention to tensions and complexities sustained within Marx's writing.

Berman argues that Marx recognized a paradoxical conjunction within capitalism between freedom and constraint. On the one hand, the relentless commodification induced by capitalism strips away all myths, mysteries and illusions, so that, 'men at last are forced to face with sober senses the real conditions of their lives and their relations with their fellow men' (Berman 1982: 95). On the other hand, the exploitative character of capitalism is deeply destructive of human potential, so that 'people can develop only in restricted and distorted ways . . . everything non-marketable gets draconically repressed or withers away for lack of use, or never has a chance to come to life at all' (ibid.: 96).

In this way, Marx shows how the working classes are caught within the structures of capitalism and how capitalism provides the means by which they might challenge these structures. He shows how the humanistic ideal of individual self-development operates ideologically to obscure the reality of class relations *and* carries within it the possibility of going beyond a capitalist social order: he views individual agency as essentially mythical and as potentially empowering. Correspondingly, Marx traces a narrow line between defining working-class consciousness as deeply damaged and as the source for revolutionary politics.

Within this kind of framework, Marx's emphasis shifts, especially between his earlier concern with the alienation of waged workers and his later concern with the exploitative but contradictory character of capitalism. But throughout, there remains an abiding tension in his view of subjectivity: in short, Marx draws upon, as well as opposes, the liberal humanist subject. This is demonstrated most clearly in his distinction between a class-in-itself and a class-for-itself. While the class-in-itself consists of decentred, alienated subjects, the class-for-itself introduces the possibility of politically salient self-awareness. This categorization implies the existence of something approximating to an authentic working-class identity or 'true' consciousness, to be distinguished from, and recovered from, the distorted, inauthentic, inadequately politicized, 'false' variety. Collective political action acts both as the medium in which class-conscious subjects can emerge and as the product of their consciousness: fostering a common identity or sense of community is intrinsic to the project of awakening a radical, proletarian subject. Thus, Marx's conception of subjectivity sits on a cusp between an anti-humanist notion of consciousness as produced by social forces beyond the control of the individual, and a more humanistic notion of the possibility of achieving (at least temporarily) a stable, coherent, common, authentic identity.

Although he traces this cusp with great subtlety in his theoretical writing, Marx did not examine how it could be negotiated in practice, and his legacy has tended to reproduce rather than resolve a polarization between structuralism and humanism. Consequently, his notion of the working classes as political agents has generally operated in conflict, rather than in harness, with his emphasis on the subject as decentred.

Moreover, although Marx's conception of the oppositional political subject is not offered as universal but as the product of a particular class position, it still claims a greater generality than is warranted: his qualified humanism reproduces some of the exclusive, minoritizing effects associated with the liberal humanism he opposed. Most obviously, Marx's proletarian subject is implicitly male.[10] This has made the relationship between Marxism and feminism as complex as that between psychoanalysis and feminism.[11]

In relation to identity politics, one important manifestation of this relates to the dichotomy between public and private. For both feminists and socialists, the private sphere can serve to conceal oppression from those who suffer it. But within class analysis, the problem is that a private domain of family life can have the effect of successfully compensating for the exploitations of waged labour, whereas within a feminist analysis, the problem is that relations of exploitation within the family remain obscured precisely because of the notion of 'home as haven'. Fostering class consciousness entails an insistence on the salience of an identity forged in the public sphere, notwithstanding its private compensations. Sometimes this has inspired a working-class politics in which any exploration of identity is anathema: in such circumstances, political loyalty and self-sacrifice are accorded high value while self-conscious examination of identity is dismissed as 'navel-gazing' (Brunt 1989). By contrast, as I elaborate in the next section, fostering feminist consciousness means bringing private experiences into the public domain, thereby creating a new kind of public identity for women. Thus, a questioning, or deconstruction, of the public/private dichotomy, especially through a problematization of identity, has the potential to divide as much as to unite.[12]

In summary, in terms of developing a politics through which to resist the universalizing claims of liberal humanism, the Freudian legacy is problematic because of its lack of commitment to an emancipatory politics, and the Marxist legacy is problematic in its attribution of a unique authenticity to the subject of a revolutionary working class. This analysis reiterates the difficulty of identity politics in terms of anti-humanism versus humanism: Freud, especially as interpreted by Lacan, leans too far in the direction of anti-humanism (and in so far as he does not, he fails to challenge existing social norms); whereas Marx leans too far in the direction of humanism (and in so far as he does not, he fails to show how identity might be political).[13] Consequently, neither legacy spawned a radical identity-based politics and both present difficulties in relation to questions of gender identity.

Of course, a contemporary politics of identity need not be bound by the limitations of these antecedents. As I have presented Marx's and Freud's ideas about subjectivity, it would appear that some *rapprochement* between the two bodies of thought might be productive. As noted above, attempts to negotiate theoretically between Marxism and psychoanalysis are not particularly promising. Therefore, in this chapter I turn instead to the political practices of second-wave feminism, which, I suggest, have marshalled (often unwittingly) aspects of both a Marxist and a Freudian tradition in the development of a politics of identity. I will show that, in so doing, feminism has retained and developed affinities with both humanism and anti-humanism.[14] In recent years, this tension has been manifest in debates about essentialism.

FEMINISM AND IDENTITY POLITICS

Consciousness-raising, which was particularly prominent in the early years of second-wave feminism, is readily interpreted in terms of the creation of a new kind of political subject. Consciousness-raising groups provided a context in which private troubles took on political meanings. Through sharing their experiences, what women had felt as personal inadequacies, neuroses and so on, came to be viewed as the product of contradictory pressures on women and dominant myths about femininity. This enabled women to rewrite their own stories, to insist that 'the personal is political', and to develop a feminist identity through which to challenge the subordination of women.

There are echoes here of Marx's understanding of class consciousness: women, like 'industrial workers gradually awaken to some sort of class [feminist] consciousness and activate themselves against the acute misery and chronic oppression in which they live' (Berman 1982: 91). From this perspective, feminist identity, like class identity, is forged through collective action: 'what we are is what we do' (Bourne 1987: 22). Indeed, despite the potential for conflict outlined above, many feminists have allied themselves both theoretically and politically with socialist politics.[15]

Consciousness-raising also has echoes of Freud's interpretation of subjectivity: the feminist argument that familiar interpretations of personal experience are misleading, and the associated search for 'deeper' meanings, have counterparts in the psychoanalytic notion of unconscious, psychical forces. Ideas of 'splitting', and of identity as fractured, resonate across the two domains. Thus, Sandra Lee Bartky describes becoming a feminist in terms of developing a 'radically altered consciousness of oneself' and a 'divided consciousness' incorporating a sense of injury and victimization on the one hand and a sense of empowerment and personal growth on the other (Bartky 1990: 12–16). But consciousness-raising, unlike psychoanalytic practice, insists at outset that the participants are political agents. In this way, feminism resists the depoliticizing tendencies of psychoanalysis.

Through consciousness-raising, therefore, the feminist movement appeared to develop a politics of identity that harnessed both an anti-humanist conception of the human subject and a commitment to political agency. On the one hand, experience was recognized as open to conflicting interpretations and subjectivity as internally divided, so that a feminist identity was necessarily constructed rather than uncovered, changeable rather than fixed. On the other hand, a sense of purpose and authorship, essential for political action, were retained through an insistence on the *validity*, rather than the transparency, of collectively articulated experiences. But, as the evolution of second-wave feminism testifies, this integration of the decentred subject and the political agent was fragile and flawed. The delicate balance of an essentially inauthentic identity politics did not hold.

To elaborate, consciousness-raising challenged the claims to universality of the subject of liberal humanism by exposing its implicit masculinity. This allowed women to reinterpret their experiences of failing to conform to such images of the subject as a failure of the model rather than of themselves. But, in developing an alternative feminist subject, some of the same limitations were reproduced (in much the same way as with the Marxist notion of a revolutionary proletarian subject). The experiences generalized and validated within consciousness-raising groups tended to be those of white, Western, middle-class, heterosexual women. Moreover, it was widely assumed that mutual identification and a sense of community were vital ingredients of feminist political practice (see Young 1990). This set up new forms of exclusiveness as alienating and oppressive to women who did not conform to this norm as either the liberal humanist subject or the authentic proletarian subject it supposedly opposed (see Spelman 1990).

I want to outline responses to the criticisms this provoked, beginning with attempts to adapt essentialism for feminist purposes and moving on to positions that radically oppose essentialism. Judith Butler summarizes these strategies as follows:

> In response to the radical exclusion of the category of women from hegemonic cultural formations on the one hand and the internal critique of the exclusionary effects of the category from within feminist discourse on the other, feminist theorists are now confronted with the problem of either redefining and expanding the category of women itself to become more inclusive (which requires also the political matter of settling who

> gets to make the designation and in the name of whom) or to challenge the place of the category as part of a feminist normative discourse.
>
> (Butler 1990: 325)

Critiques of the exclusiveness of the women's movement prompted the recognition and development of *multiple* feminist identities through a process of hyphenation: women began to identify as black-feminist, working-class-feminist, lesbian-feminist, Jewish-feminist, and so on. As with early feminist consciousness-raising endeavours, groups that formed around these labels sought, through the sharing of experiences, to understand the nature of their oppression in a manner conducive to resistance. Acknowledging these kinds of differences became important in other groups too. For example, Mary Louise Adams describes an 'exercise' undertaken by members of a women's crisis centre in which

> each woman spoke in turn: 'I am a white, working-class, heterosexual woman'; 'I am a white, middle-class, lesbian Jew' . . . The more politically astute would add further categories – 'I am a white, middle-class, ablebodied, Anglophone lesbian' – and the facilitator would nod her approval at each innovation. . . . The idea was that an understanding of various oppressions and how we were affected by them or colluded with them would make us better counsellors and advocates to the women who sought our services.
>
> (Adams 1989: 22)

Despite these intentions, the development of this style of identity politics has been problematic for the women's movement. While it challenged misguided assumptions about 'sisterhood' in important and necessary ways, the assertion of multiple identities eclipsed the earlier emphasis on identity as fractured. Reliance upon apparently pre-given categories of class, sexual orientation, race, ethnicity and so on invoked a conception of identity as something to be acknowledged or uncovered rather than constructed, as something fixed rather than changing. Implicitly, therefore, this approach to issues of the differences among women appealed to a conception of the subject as centred and coherent, that is, to something very close to a liberal humanist model. The chief difference lay in the insistence that there are many, rather than just one, essences of identity.

Some of the consequences of this style of identity politics have been divisive. In conjunction with the importance attached to mutual identification, it encouraged what Donna Haraway describes as a

> Painful fragmentation among feminists (not to mention among women) along every possible fault line. . . . The recent history for much . . . feminism has been a response to this kind of crisis by endless splitting and searches for a new essential unity.
>
> (Haraway 1990: 197)

More insidiously (and again echoing a tendency within socialist politics), it encouraged an equation of oppression with authentic knowledge and even virtue, and of privilege with taintedness. Those positioned lower down within a hierarchy of oppressions were assumed to have greater insight than those higher up. This inevitably accentuated fragmentation and detracted from the emancipatory goals of feminism. It prompted Jenny Bourne's negative appraisal of identity politics cited at the beginning of this chapter (p. 84).

The second response to the exclusions of feminism embraced the post-structuralist critique of liberal humanism and sought to forge a postmodern feminism. Within this perspective, anti-essentialism is a powerful motif and identity is conceptualized as fluid and malleable. Consequently, any attempt to ground politics in identity becomes deeply problematic.

Although, as I will illustrate later in this chapter, feminism in particular and emancipatory politics in general can deploy post-structuralism to good effect, there are great dangers associated with this position. Taken to its logical conclusion, the category 'women', upon which feminism is based, becomes a free-floating sign apparently able to take on any meaning we give it. The materiality of social relations is, in effect, ignored, in favour of a domain of representation in which stuctures of power are treated as illusory. As Linda Alcoff notes, the effect is remarkably similar to that of liberal humanism:

> For the liberal, race, class and gender are ultimately irrelevant to questions of justice and truth because 'underneath we are all the same'. For the post-structuralist, race, class, and gender are constructs and, therefore, incapable of decisively validating conceptions of justice and truth because underneath there lies no natural core to build on or liberate or maximize. Hence, once again, underneath we are all the same.
>
> (Alcoff 1988: 421)

Such a position is post-feminist rather than feminist, and is inimical not only to identity politics but to any notion of emancipatory goals and to any form of collective politics. It is a position according to which 'there is no extra-discursive justification of values, no transcendent "rights" and "wrongs" to which we can appeal in our denunciations of "oppression" and aspirations to "emancipation" ' (Soper 1991: 99). Clearly, this formulation relinquishes political agency in favour of the notion of the subject as decentred and incoherent. It carries resonances of the Lacanian account of subjectivity in its abandonment of emancipatory goals.

Thus, the evolution of feminist political practice brings the analysis back to the tension between anti-humanism and humanism: the strategies outlined are flawed because they lean too far towards one or the other. These difficulties have provoked a re-examination of some of the basic precepts of feminism. In particular, both the status of experience and the nature of identity have come in for close scrutiny. In the interpretations I advance here, my intention is to retain the emancipatory potential of critiques of liberal humanism without succumbing to the apolitical relativism of postmodernism.

Personal experience has been a key term within feminist politics. It provided the material from which a decentralized, non-hierarchical feminist politics was constructed. Consciousness-raising was all about women sharing their experiences of being women. The identity politics of what I have termed 'hyphenated feminisms' was all about women sharing their experiences of being women of particular classes, races, ethnicities, sexualities and so on. This was crucial in order to challenge dominant perceptions of women's experience. The flaw was to remain too close to liberal humanism by assuming that knowledge flowed directly from experience and that experience ensured the authenticity of knowledge. This implies that, rather than being constructed, experience has the quality of an irreducible essence, which resides in such characteristics as female-ness, middle-class-ness, white-ness and so on.[16] It also invokes a kind of personal immunity in that to authenticate knowledge in terms of personal experience is to make one's ideas and one's being indistinguishable. Consequently, anyone who criticizes knowledge generated in this way is liable to be accused of attacking the person from whom it originated.

While I would argue that attempts to use experience to authenticate knowledge in this way should be rejected, it is, surely, important to retain some sense of experience as valid. Without this, there is little scope for resisting dominant groups' oppressive definitions of subordinated others. This is the problem that arises with anti-humanism. But, in this context, my choice of terms is advised: to claim that experience is valid is not the same as claiming it to be true; rather, it allows experience to be understood as a salient but contestable, rather than as a foundational, phenomenon. It acknowledges that 'experience is a difficult, ambiguous and often oversimplified term'(de Lauretis 1990: 261). From this perspective, we each inhabit our own experientially valid realities, and as social beings we negotiate between our partially overlapping, and existentially vital versions. The essence of experience, therefore, resides not in its authenticity but in its ability to inform our ideas and actions.[17] And feminist theory enters into a recursive relationship with articulations of experience:

> For a theoretical perspective to be politically useful to feminists, it should be able to recognize the importance of the *subjective* in constituting the meaning of women's lived reality. It should not deny subjective experience, since the ways in which people make sense of their lives is a necessary starting point for understanding how power relations structure society. Theory must be able to address women's experience by showing where it comes from and how it relates to material social practices and the power relations which structure them. It must be able to account for competing subjective realities and demonstrate the social interests on behalf of which they work.
>
> (Weedon 1987: 8)

In this way, it becomes possible to negotiate between the competing claims of humanism and anti-humanism.

Turning to the question of identity, I think it is important to distinguish between fractured subjectivities and multiple identities. The notion of multiple identities associated with 'hyphenated feminisms' treats identity as something to be discovered, reclaimed and celebrated, rather than created. This is to be achieved by exploring one's experience of oppression, which becomes an act of resistance in itself. It invokes a concept of separate identities, each of which is fixed and stable. In contrast to this, the anti-humanist position, which I have associated with an emancipatory politics of identity, views subjectivity as fractured and decentred. Singular, coherent identities are only ever mythical constructs. This tendency, associated with anti-humanism, needs to be tempered by a recognition that the ability to differentiate between different constructions of identity – to accept that some are more significant than others, that some are expressions of 'poles of world historical systems of domination' (Haraway 1990: 203) – is politically essential.

In this context, identity politics is about deconstructing *and reconstructing* (necessarily multiple) identities in order to resist and undermine dominant mythologies that serve to sustain particular systems of power relations. Jane Gallop expresses it thus:[18]

> I do not believe in some 'new identity' which would be adequate and authentic. But I do not seek some sort of liberation from identity. That would lead only to another form of paralysis – the oceanic passivity of undifferentiation. Identity must be continually assumed and immediately called into question.
>
> (Gallop 1982: xii)

This conceptualization seeks to avoid the essentialism implicit in appeals to authentic identities while acknowledging that we cannot do without identity altogether. It defines identity not in the realm of real essence, nor in the realm of a received mythology, but in the realm of a context-dependent creativity. In other words, fictions of identity are essential, and essentialism (humanism) is deployed strategically rather than ontologically.

This interpretation of identity is perhaps best illustrated within contemporary sexual politics:

> The 'sexual outlaws' [such as male homosexuals] have disrupted the categorisations of the received texts and have become thinking, acting, living subjects in the historical process. The implication of this is that . . . modern gay identities . . . are today as much *political* as personal or social identities. They make a statement about the existing divisions between permissible and tabooed behaviour and propose their alteration. These new political subjectivities above all represent an affirmation of homosexuality, for by their very existence they assert the validity of a

> particular sexuality. This surely is the only possible meaning of the early gay liberation idea of 'coming out' as homosexual, of declaring one's homosexuality as a way of validating it in a hostile society. Arguments that this merely confirms pre-existing categories miss the point. The meanings of these negative definitions are transformed by the new, positive definitions infusing them. The result is that homosexuality has a meaning over and above the experience of a minority. By its existence the new gay consciousness challenges the oppressive representations of homosexuality and underlines the possibilities for all of different ways of living sexuality. This is the challenge posed by the modern gay identity. It subverts the absolutism of the sexual tradition. . . . These identities [gay and lesbian] are not expressions of secret essences. They are self-creations, but they are creations on ground not freely chosen but laid out by history.
>
> (Weeks 1985: 200–1, 209)

More succinctly, it is captured in the slogan: 'I am out therefore I am' (see Sedgewick 1991). And this conceptualization of identity underpins attempts by feminists to come to terms with the exclusionary effects of an insufficient sensitivity to differences among women without either appealing to ontological essences or relinquishing completely the identifier 'women' on which feminism is based.[19]

PLACE AND ESSENCE

This kind of politics of identity draws on ideas implicit in early consciousness-raising groups, but goes further in making explicit the notion of identity as process, as performance, and as provisional. In so doing, it acknowledges that identity is always both internally fractured and externally multiple. The possibility of doing this owes a good deal to post-structuralism and its associated vocabulary. Thus, while I began this chapter by locating identity politics in relation to Marx and Freud, I want to close it by moving to a different place.

Post-structuralism has relied strongly on spatial terms of reference, and the reconceptualization of identity politics I have drawn out effectively spatializes our understanding of familiar categories of identity like class, nationality, ethnicity, gender and so on. Rather than being irreducible essences, these categories become positions we assume or are assigned to:

> the poststructuralists' deconstruction of the subject-as-agent allowed an understanding of the subject as a position within a particular discourse [fiction]. This meant that a subject was no longer coterminous with the individual. Rather, the power-knowledge relations which produced a subject-position implied that there was no necessary coherence to the

> multiple sites in which subject positions were produced, and that these positions might themselves be contradictory.
>
> (Henriques, Hollway, Urwin, Venn and Walkerdine 1984: 203)

The metaphor of position is deployed to capture both the multiplicity and internal fracturing of identities, while the concept of subject reminds that we still operate with narratives of our individual integrity.

One consequence of the spatial metaphor is that the question 'Who am I?' evident in some versions of identity politics becomes 'Where am I?' In this way, place takes the place of essence. But I would argue that it does not banish essentialism. Rather, I would follow Diana Fuss who deconstructs the opposition between essentialism and constructivism, and who suggests that, 'the essentialism of "anti-essentialism" inheres in the notion of place or positionality' (Fuss 1989: 29). She continues, 'what is *essential* to social constructionism is precisely this notion of "where I stand", of what has come to be called, appropriately enough, "subject-position" ' (ibid.).

This account implies that geographical terms of reference do the work done by essences in other formulations. It turns a politics of identity into a politics of location (see Probyn 1990). But, while I am suggesting that place is in effect a repository for the essence identity politics cannot wholly do without, I am not suggesting that this substitution adds nothing to the practice of identity politics. Rather, the question becomes one of how place and related spatial metaphors are understood. If their meanings are left unexplored, the effects are likely to be as undermining as an inadequately conceptualized notion of identity. More specifically, if references to 'place', 'position', 'location' and so on covertly appeal to fixed and stable essences, rather less will have been gained than we might hope. And this is likely to be the case in so far as these metaphors import a Cartesian conceptualization of space as an absolute, three-dimensional grid devoid of material content. Writing of an apparent 'reassertion space in social theory', Neil Smith diagnoses precisely this usage:

> It is to this conception of space as ground . . . that spatial metaphors invariably appeal . . . Whatever the power of spatial metaphors to reveal especially the fragmented unity of the contemporary world, they work precisely by reinforcing the deadness of space and therefore denying us the spatial concepts appropriate to *analysing* the world.
>
> (Smith 1990: 169)

I am less pessimistic than Neil Smith. It seems to me that the emphasis on *where* – on position, on location – is allowing questions of identity to be thought in different ways. For example, these metaphors appear to be encouraging a concern with the relationships between different kinds of identities and therefore with the development of a politics grounded in affinities and coalitions, rather than some pristine, coherent consciousness.[20] This move is likely to allow a politics of identity to negotiate more effectively between the opposing pulls of essentialism and anti-essentialism. I would therefore endorse the claim that geography has an important contribution to make to contemporary politics. The point is that if they are to retain their potency, the geographical metaphors of contemporary politics must be informed by conceptions of space that recognize place, position, location and so on as *created*, as *produced*.[21] Then the positionality of identities can be deployed imaginatively and creatively to construct a politics that, in Donna Haraway's words, can 'embrace partial, contradictory permanently unclosed constructions of personal and collective selves and still be faithful [and] effective' (Haraway 1990: 199).

ACKNOWLEDGEMENTS

Thanks to Erica Burman and Hazel Christie for their comments on an earlier version of this chapter.

NOTES

1 See also Adams (1989).
2 More specifically I found myself deeply uneasy about the celebrations of identity that seemed to me to be almost inescapable in the women's movement in Britain during the early to mid-1980s. For example, like Lynn Segal (1986) I found much of the imagery of feminine virtuosity (and especially maternity) associated with the mass protests at Greenham Common personally and politically alienating. And I was also disturbed by a tendency for *questions* about sexual orientation to be eclipsed by an apparent preoccupation with *categories*.
3 My account is inspired by Grosz (1990: 1–2).
4 Terms such as 'liberal humanism' carry complex meanings and references. These are easily and too frequently reduced to one dimension that amounts to little more than a caricature. Of the potential candidates for such treatment in this essay, I suspect I come closest to such an abuse in my use of the term 'liberal humanism'. While I attempt to sustain a more nuanced and variegated use of the term 'humanism', 'liberal humanism' is, perhaps, my bogeyman. The difficulty arises in part from my decision to write about identity politics within a framework of conceptual dichotomies. While in sympathy with attempts to move beyond either/or to both/and formulations (itself an either/or formulation), my principal concern in this essay is with understanding how the either/or structure operates in identity politics with respect to a humanism/anti-humanism dichotomy. In so doing, I do not claim to have worked through the dichotomy adequately: I continue to reproduce it even as I seek to unsettle it.
5 See, for example, Hartsock (1987).
6 On the universalizing/minoritizing dualism, see Sedgewick (1991).
7 See Turkle (1979); Henriques, Hollway, Urwin, Venn and Walkerdine (1984).
8 Also see Mitchell (1974); Craib (1990).
9 Discussed, for example, by Mitchell (1974); Weeks (1985); Craib (1989).
10 See, for example, Hartmann (1981); Sydie (1987).
11 See Rowbotham, Segal and Wainwright (1980); Hartmann (1981).
12 See, for example, Pateman (1989).
13 Freud's commitment to scientific objectivity with its implicit Cartesian subject,

and Marx's commitment to an authentic subjectivity, have led to both being classified as archetypal modernists. But as I have emphasized other aspects of their respective projects suggest that they were precursors of postmodernism. Perhaps the biggest problem lies in the conceptualization of a modernism/postmodernism break. While I do not explore this dualism explicitly here, I hope at least to contribute to a questioning of its oppositional structure.

14 See Soper (1990).
15 See, for example, Bartky (1990); Segal (1991).
16 See Fuss (1989: 25).
17 See Fuss (1989) who draws on Locke's distinction between real and nominal essences (especially pp. 4–5).
18 See also Young (1990).
19 See, for example, Alcoff (1988); Butler (1990); Flax (1990).
20 See, for example, Haraway (1990).
21 See Lefebvre (1991).

REFERENCES

Adams, M. L. (1989) 'There's no place like home: on the place of identity in feminist politics', *Feminist Review* 31: 22–33.

Alcoff, L. (1988) 'Cultural feminism versus post-structuralism: the identity crisis in feminist theory', *Signs* 13: 405–36.

Bartky, S. L. (1990) *Femininity and Domination*, New York and London: Routledge.

Berman, M. (1982) *All That Is Solid Melts Into Air*, London: Verso.

Bourne, J. (1987) 'Homelands of the mind: Jewish feminism and identity politics' *Race and Class* 29: 1–24.

Brunt, R. (1989) 'The politics of identity', in S. Hall and M. Jacques (eds) *New Times. The Changing Face of Politics in the 1990s*, London: Lawrence and Wishart, 150–9.

Butler, J. (1990) 'Gender trouble, feminist theory and psychoanalytic discourse', in L. Nicholson (ed.) *Feminism/Postmodernism*, New York: Routledge, 324–40.

Craib, I. (1989) *Psychoanalysis and Social Theory*, Amherst: University of Massachusetts Press.

de Lauretis, T. (1990) 'Upping the anti [*sic*] in feminist theory', in M. Hirsch and E. F. Keller (eds) *Conflicts in Feminism*, New York: Routledge, 255–70.

Flax, J. (1990) *Thinking Fragments. Psychoanalysis, Feminism and Postmodernism in the Contemporary West*, Berkeley, CA: University of California Press.

Fuss, D. (1989) *Essentially Speaking*, London and New York: Routledge.

Gallop, J. (1982) *The Daughter's Seduction. Feminism and Psychoanalysis*, Ithaca, NY: Cornell University Press.

Grosz, E. (1990) *Jacques Lacan. A Feminist Introduction*, London and New York: Routledge.

Haraway, D. (1990) 'A manifesto for cyborgs: science, technology, and socialist feminism in the 1980s', in L. Nicholson (ed.) *Feminism/Postmodernism*, New York: Routledge, 190–233.

Hartmann, H. (1981) 'The unhappy marriage of Marxism and feminism: towards a more progressive union', in L. Sargent (ed.) *Women and Revolution*, London: Pluto.

Hartsock, N. (1987) 'Rethinking modernism: minority vs. majority theories', *Cultural Critique* 7: 187–206.

Henriques, J., Hollway, W., Urwin, C., Venn, C. and Walkerdine, V. (1984) *Changing the Subject*, London and New York: Methuen.

Kovel, J. (1988) *The Radical Spirit*, London: Free Association Books.

Lefebvre, H. (1991) *The Production of Space*, Oxford: Blackwell.

Mitchell, J. (1974) *Psychoanalysis and Feminism*, Harmondsworth: Penguin.
Pateman, C. (1989) *The Disorder of Women*, Cambridge: Polity Press.
Probyn, E. (1990) 'Travels in the postmodern: making sense of the local', in L. Nicholson (ed.) *Feminism/Postmodernism*, New York: Routledge, 176–89.
Rowbotham, S., Segal, L. and Wainwright, H. (1980) *Beyond the Fragments*, London: Pluto.
Rose, J. (1986) *Sexuality in the Field of Vision*, London: Verso.
Sedgewick, E. K. (1991) *The Epistemology of the Closet*, Hemel Hempstead: Harvester Wheatsheaf.
Segal, L. (1986) *Is the Future Female?*, London: Virago.
—— (1991) 'Whose left? Socialism, feminism and the future', *New Left Review* 185: 81–91.
Smith, N. (1990) *Uneven Development*, (2nd edn) Oxford: Blackwell.
Soper, K. (1990) 'Feminism, humanism and postmodernism', *Radical Philosophy* 55: 11–17.
—— (1991) 'Postmodernism and its discontents', *Feminist Review* 39: 97–108.
Spelman, E. V. (1990) *Inessential Woman*, London: The Women's Press.
Sydie, R. A. (1987) *Natural Women and Cultured Men*, Milton Keynes: Open University Press.
Turkle, S. (1979) *Psychoanalytic Politics*, London: Burnett Books.
Weedon, C. (1987) *Feminist Practice and Post-structuralist Theory*, Oxford: Blackwell.
Weeks, J. (1985) *Sexuality and its Discontents*, London: Routledge and Kegan Paul.
Wilson, E. (1987) 'Psychoanalysis: psychic law and order?', in Feminist Review (ed.) *Sexuality: a Reader*, London: Virago, 157–76.
Young, I. M. (1990) 'The ideal of community and the politics of difference', in L. Nicholson (ed.) *Feminism/Postmodernism*, New York: Routledge, 300–23.

6

WOMEN'S PLACE/EL LUGAR DE MUJERES

Latin America and the politics of gender identity

Sarah A. Radcliffe

> . . . the boots, the slaps and the kicks were necessary before we good housewives would finally go out and participate and . . . produce a show of protest. So I'm angry for not having left my knitting and pots and pans earlier to go out and complain about the tanks.
>
> (Bonafini 1985: 73)

With its light irony and homely references, the quote above refers to a wide-reaching political juncture which unified numerous women throughout Latin America during the 1970s. During these years, military governments took over their countries and, in many cases, systematically abused human rights and reduced the spaces for civilian politics. Hebe de Bonafini's testimony is one of the many produced by Argentine women whose daughters and sons were 'disappeared' by the military during the dictatorship of the late 1970s (cf. Sternbach 1991). As the military reached into private homes to extract civilians (and later kill them in many cases),[1] so women entered into new locations (the streets and main squares) and different political positions in the changed political context, as Hebe describes above. What did these dislocations and repoliticizations mean for women's identities during these years? What role did the geography of gender relations, in which public and private had been so clearly separated in space and meanings throughout Argentinian history, play in shaping the new forms and discourses of female political identities? It is with these questions in mind that this chapter examines recent gender politics of Argentina, in the light of general Latin American experiences of women's mobilization, political instability and economic crisis. In this chapter, the multiple social and geographical spaces inhabited by women in Latin America, as well as the meanings of these spaces, are discussed. In particular, my focus will be on the shifts in locations, meanings and identities of women during the military dictatorship in Argentina during the period 1976–83.

The importance of this case study lies with the symbolism which has grown up around it, and the metaphors around space (particularly metaphors of public and private) used in analysis. It has often been claimed in relation to the political mobilizations of women under the military in Argentina that the only political space 'left open' was that for previously 'apolitical' women, and that as a function of their (deliberately) depoliticized yet increasingly organized movement, women gained new identities and formed a new community of resistance in the country, which in turn shaped the nature of the return to democracy.[2] For example, in her review of Latin American women's movements, Jane Jaquette argues that with the change from a private to a public sphere of activities, women gained a new politics, differentiated only in their methods of 'gaining access to the public sphere' (Jaquette 1989: 188). Suggesting that the emergence into politics breaks the barriers between 'private' and 'public', this process leads to women being 'the subjects, not merely the objects, of political action' (ibid.: 189). Such conflations of place, metaphor, subjecthood and communities of political activists are found in other analyses of women's movements in Latin America during the past few decades. Sonia Alvarez's excellent discussion of regime transitions in Brazil points to how women's movements 'constitute deliberate attempts to push, redefine, or reconstitute the boundary between the public and the private, the political and the personal' (Alvarez 1990: 23). In his more cursory discussion of the Madres of Argentina, Michael Taussig argues that women take from the state its metaphor of women-as-nation, due to their emergence as political subjects (Taussig 1990: 210). Other examples could be cited.[3]

There are several problems with this analysis of women's movements in Latin America, not least in relation to Argentina. Three points can be made here. First, the terms 'public' and 'private' remain largely unexamined in relation to the distinctions between spatial and metaphorical aspects: the two dimensions are conflated. Second, separating out public and private as two dichotomous spaces (whether in geographical or metaphorical terms) reproduces socio-cultural categories rather than opening them up for analysis of the inter-linkages and interrelationships which between them give rise to this duality. And finally, the relationship between spaces (public–private/spatial–metaphorical) and identities is unclear: subjecthood is seen to adhere to 'public' spheres, objecthood to 'private', yet the mechanisms of transformation of identities, and the linkages between these and spaces of operation for politics remain unexamined. It is these three aspects, rather than a straightforward 'history' of the Argentinian women's movement, that forms the content of this chapter. My starting point is that the reality of women's lives go beyond simple dichotomies, and are embedded in active engagement with subjecthood, identity and social transformation, whatever their political positions.

Recent feminist theory has highlighted how women are active agents in constructing their own identities. Through the life experiences and the discourses into which they are inserted, female subjects are formed and reproduced (de Lauretis 1987). Breaking from the deep structures of metaphor in which (abstract) space is associated with femininity and passivity and (abstract) time connotes masculinity and action (Massey, Chapter 8 of this volume), it is also important to recognize and identify situations in which female subjects use history and movement through space to construct their own (personal/political) identities. In other words, the (self-) representation of gendered identity is evidenced by the interrelationships of place and history, their associational meanings, and gendered positionings in relation to these abstracts. In order to examine these issues, careful contextualizations of meaning and actions are necessary, given the geographical variations in socio-cultural 'technologies of gender'.[4] In this respect, the political meanings of gender-specific movements are not easily categorized as 'progressive' or 'feminist': the politics around gender as a mobilizing centre for identities depends on the histories of gender relations and wider socio-political power relations, which vary so widely from one place to another.

In Latin America, to see women as agents in identity politics is to challenge the hegemonic structures of Latin American society around the family, sexuality, Catholicism and formal politics, which assign women to an unproblematized position as family maintainers on a daily and generational basis (Cubbitt 1988). Although largely a secular society now, Argentina's history and gender relations were informed by the Catholic notions of female purity and sacrifice to a degree not found in other Latin American countries where syncretism between indigenous and Catholic belief is more widespread.[5] Right-wing Catholicism informed the emergence of a conservative nationalism, in which notions of heirarchy and vertical structures of authority were re-expressed throughout the mobilizing years of president Juan Perón (1946–55, 1973–4)[6] (Calvert and Calvert 1989). Argentinian family structure, as in other Latin American countries, is profoundly influenced by such ideas and explains the permanence of notable family networks in the country, in which women act as marriage pawns between powerful patriarchical units (Jelin 1991). Reinforced by racist ideologies of European heritage and non-black/indigenous pasts, families in Argentina focus significant discourses around duty, favouritism and loyalty (Calvert and Calvert 1989: 140). Within this context, legal provisions place strict controls over women's lives via their husbands' authority. Through the twentieth century, for example, husbands have decided on the management of property, the location of the marital home, and whether wives could work. In Argentina too, women's legal status changes on marriage and full rights over children were granted to husbands under the *patria potestas* law. Upon divorce, the court could place women in an 'honest house'. Moreover, the rape of married women was considered more of a crime than that of unmarried victims (Medina 1989).

Gender relations are further shaped by the cult of masculinity around the gauchos, mythical founding fathers of the Argentine nation (Sarmiento 1845;

also Rowe and Schelling 1991). Living on the wild plains, the gauchos, according to environmentally determinist discourses around gender and nationalism in Argentina, were disdainful of family life and women, for whom they had only contempt.[7] Despite the early emergence of feminism in Argentina (Carlson 1988), women's lives there remain more closely bound to the home and domestic duties than is true for women in comparably industrialized countries in the continent, such as Brazil and Chile. Female labour force participation remains low, although it increased during the economic crisis of the 1980s which prompted female paid-work participation. However, women continued to be viewed as the primary domestic workers, resulting in a 'double day' of work. One survey found that women did 90 hours' work per week, compared with an average of 40.3 hours by men (Calvert and Calvert 1989: 143).

Given the domestic and nurturing role that women have historically been assigned to in practice and discourse in Argentina, it is not surprising that political participation by women has been limited.[8] Women's right to vote was granted in 1947, Argentina being one of the last countries in Latin America to introduce this (Jaquette 1989: 3).[9] While in most Latin American countries women were granted the vote on the expectation of their conservatism and support for authoritarian regimes,[10] women's support for the populist and corporatist regime of Juan Perón and his first wife Eva Perón was organized throughout the 1940s by means of women's party sections, improved conditions for female workers, and encouragement of female public officials (Jaquette 1989: 4; Chester 1986). By the early 1970s, the mobilization of women was no longer a priority, as the right-wing inheritors of Perón's mantle struggled with economic crisis, political violence and military unrest. Various anti-women measures were introduced, curbing female liberty in interpersonal relations through restricting contraceptive availability and female rights over children (Feijoó 1989). Feminist organizations, although growing in numbers and strength, remained insignificant in the wider political scene – because of their own concerns as much as the growing political crisis (Chester 1986).

This political crisis came to a head in March 1976 when a military government took power, in response to what it perceived as a breakdown in public order, and attacks on an outpost of 'Western civilization'. Rapidly moving to gain control of the country, the military instituted a series of neo-liberal economic policies in an attempt to reduce inflation and generate economic growth, as well as acting to quell any opposition. Repressive military methods including detention, disappearance and torture were initiated by the military and applied to almost the entire population, including women (Fletcher 1992).

The context for such a repressive response and the severe curtailment of civilian political space, must be understood in terms of Argentinian history. Military regimes were regular features of Latin American early republican history during the nineteenth century, as strongmen *caudillos* fought over regions and populations. During the twentieth century, generals intervened in favour of 'nationalist' goals of stability and growth, in which human rights were often subordinated to 'order and progress'. In Argentina, the fissures and distrust between different groups in society, in part exacerbated by the familial structure described above, highlighted the difficulty of distinguishing between enemies and friends. As a consequence, the enemies of the nation could include civilian populations who disagreed with the nationalism espoused by the military, which highlighted 'Western and Christian values' (Diago 1988: 25; Calvert and Calvert 1989: 10; Westwood 1989). Thus in Argentina, detentions and disappearances were utilized to rid the country of 'subversives', a term which included the guerrilla movements of the Montoneros and the People's Revolutionary Army (ERP), other political activists, leftist intellectuals and relatives. Such moves were widely supported by the Catholic hierarchy, whose conflation of patriotism, service to the nation, and Argentinian safety was backed by religious doctrine.

The gender politics of the military regime installed in 1976 emphasized a mythical return to the family as the 'basic cell of society' where all other associative links would be broken (Feijoó and Gogna 1990: 84; Feijoó 1989: 75). Within this model of society, women were to retain order by overseeing children's behaviour while their domestic roles as wives and mothers were romanticized. Women were not expected to be politically informed about events, but were to maintain a stable base within the home through their 'knitting and pots and pans', as described by Hebe de Bonafini (Bonafini 1985). Their official identity was that of a negated political subject, the purest safe apolitical community, embedded within the dangerous(ly) political public world. For women under military regimes elsewhere in Latin America faced with this situation, it was a community of a 'denied identity, not constituted' (Kirkwood 1988: 20).

Nevertheless, in practical terms, women experienced the political world at first hand. Thirty per cent of those registered as arrested or as having disappeared were women (cited in Feijoó and Gogna 1990: 84). Women taken into the clandestine detention centres which sprang up around the country were subject to rape and sexual abuse, regardless of the reason for arrest (Fletcher 1992: 10). Women were separated from their children born in detention who were subsequently adopted, often by families who had no knowledge of the reasons for the parents' absence (*Mujer/Fempress* no. 85, 1988).

However, despite this gender-specific treatment of violence and the rhetoric of women being above violence, women were among the first in challenging military actions and the suspension of civil rights. In Argentina, as in Uruguay and Chile, which also at that time experienced 'bureaucratic authoritarian regimes' (O'Donnell 1973),

> women were among the first to protest against the mass imprisonments and disappearances; organizations based on women relatives of the

> disappeared formed the backbone of human rights groups, and human rights became a central issue in the civilian effort to push the military out.
>
> (Jaquette 1989: 4)

How did this occur? What type of identity politics was constituted by women in this situation, where formal politics had gone underground, and the merest hint of a subversive identity could land someone in gaol?

The female community of resistance to the military regime developed within the specific geographical sites in which the military regime was active in exerting physical control and maintaining ideological hegemony. In other words, female mobilization took place in and around particular places which became significant (in real and ideological terms) under the military regime: places which had been outside the frames of reference of political meaning previously. It was in these places that Argentinian women created and reproduced a politics of opposition. Mothers encountered each other in the offices and department buildings of the security forces, in the courts and the police commissioners, in hospitals and prisons (Calvera 1990: 102). As one woman recounted,

> I went to the Caseros prison, to the old Women's Prison in La Plata. I also went over to the regiments. I was with the 1st Regiment, and from there they sent me to talk to Coronel Minicucci, in La Tablada, who wouldn't see me.
>
> (Oria 1987: 50)

Once the Madres became known to the military, they became subject to arrests, surveillance and intimidation. The first Madres' leader was arrested and 'disappeared', to be replaced by Hebe de Bonafini. To avoid police action, the Mothers acted as 'ordinary women': they met in cafés with packages of shopping, in public gardens with young children in tow, and initially in churches over prayers (Oria 1987: 112–14).

In terms of public meetings, the most significant of these places was the main square at the centre of Buenos Aires, the Plaza de Mayo (May Square). Starting in April 1977, mothers of women and men who had disappeared paraded around the central pyramid of the Square every Thursday, wearing white headscarves and at times carrying photographs or silhouettes of their relatives.[11] The group soon became known as the Mothers of the Plaza de Mayo (*Madres de Plaza de Mayo*), growing from the original group of 14 women to 150 by July of the same year. Their basic demand was the return of living relatives. Coming from diverse class backgrounds and from different parts of the city, the Madres had met during their fruitless journeys around police stations. Once their organization had been formalized in August 1979, the Madres gained the critical attention of military and public commentators, who called them 'the mad women'. Certain leaders in the Madres organization were arrested and interrogated, yet the movement gained strength throughout the national territory. By 1981, the Madres had mobilized a national movement, gaining explicit support from public figures and galvanizing a 'March of Resistance' in 1982 which brought 5,000 people out on to the streets of Buenos Aires. In the wake of the Madres group, organizations of Grandmothers (*Abuelas de Plaza de Mayo*, founded in 1977)[12] and of Relatives were created, yet the Madres group retained its original membership and women-only rules.

The gender politics of the Madres movement was clear from the start. Women explained that their husbands were at work and did not have time to attend the detention stations or the march around the Plaza. Given a predominant socio-cultural expectation of male soldiers' behaviour too, it was felt that women would not be attacked as violently as would men. In dealings with the police, female relatives seemed to get more sympathetic and direct treatment than did fathers and brothers of the disappeared. Thus, in the spaces of repression, women were symbols of vulnerability (not generally to be mistreated physically) and the major domestic family actors: these meanings arose out of the long-term cultural history of the country, as well as out of the immediate gender-politics of the military regime, which had emphasized the role of mothers in invigilating for/over their children.

Of course, the Mothers' community of identity developed in opposition to the state's treatment of their children, and not in support of the state's rhetoric about the family. However, in the imagery, practices and assumptions about the content of and the boundaries between 'public' and 'private', certain continuities exist between the military's 'basic cell' of the family and the Madres' activism. Thus, the link between mothers and children was perceived to be a direct, unmediated one begun at birth. According to the Madres, 'this chain is unbreakable – children, mothers, children – because there will always be a mother who, after giving birth to a child, fights for his/her liberty' (Madres de Plaza de Mayo 1987: 9). In this discursive relationship, mothers felt the same as their children, as if they were the same body: 'The anxious sleeplessness of each one of us Madres is because we suffer biologically and psychologically the hurt of our children in our own flesh' (Madres de Plaza de Mayo 1987: 8).

As mentioned above, women came to share among themselves knowledge and experiences through their visits to countless police and detention centres searching for relatives. These experiences, confirming and providing evidence for an unofficial truth, made the women involved aware of a potential common identity, through the sharing of spaces and histories. Without ascribing essentialisms to the people or places involved in this political situation, it is important to recognize that the relationship between people and place was one articulated through layers of interpersonal behaviour, histories of gender relations, fear and the logistics of organization. However, in this context, a collective identity of resistance could potentially have taken other paths, such as an assertion of citizenship or an identity as fighters for democracy, in which both male and female actors participated. In the event, activists opted for an essentializing notion of kinship, motherhood and life, which on the one hand turned back the military's rhetoric on itself, but on the other reinscribed an essentializing tendency of gender politics among the Madres themselves. Rejecting feminism[13] and retaining their Catholic faith,[14] the Madres created a resistance movement in which gender identities mirrored their historical context and existing power relations.

However, this did not prevent the Mothers from innovating political spaces and legitimating activites. Having lived through an apoliticizing socialization process, women were generally inexperienced in formal politics, while their knowledge of political symbolism affected their choice of site for political activism. Disregarding advice from political activists, the Madres made straight for the heart of the regime's power – the Plaza de Mayo where their concerns became visible to the regime and the public. The Madres saw the space inhabited by the Square and the institutions that surrounded it as a symbol of the nation, the President and the resolution of major problems – a space which did not belong to them (Oria 1987: 67). Despite male scepticism about the effectiveness of such a move, women believed that this location would serve their purposes of directly confronting the military junta.

As in other situations in Latin America, the Madres were concerned to highlight ethical and moral concerns in their mobilization (cf. Chaney 1979), saying that they were 'not passive but pacifist' (Feijoó and Gogna 1987: 152). Using white headscarves as life and peace symbols, the Mothers also emphasized the non-partisan nature of their demands, within the context of military repression of organized opposition and as a mechanism to distance themselves from distrusted traditional political parties. Ironically, the military regime restricted scope for civilian politics, yet mobilized those who had never experienced full citizenship rights (Jaquette 1989: 5). Being 'outside' the traditional political system, the Madres laid claim to a purity and originality of purpose, in their own politicized discourse. In common with certain women elsewhere in Latin America, the Madres distanced themselves from old-style politics while highlighting the newness of their activities, which they saw as learning experiences.[15]

In choosing to parade publicly with a maternal identity, it has been widely suggested that the Madres overturned the geography of gender in Argentina. Women moved from their traditional place in the private home, symbolic of their historical political role, into the space occupied by the most potent symbol of the military regime. As argued convincingly by Sternbach, the Mothers of Plaza de Mayo 'appropriate[d] the Plaza de Mayo both physically and symbolically and, in so doing, transgress[ed] the previously delineated female space in the name of . . . disappeared relatives' (Sternbach 1991: 97).

However, this simple reversal must be examined more carefully: the physical movement from home to street did not (and does not necessarily) translate into a symbolic power shift. In the context of the Argentina of 1977, the Madres appropriated the space of the Square within a particular discourse of identity, which focused on motherhood and life. Within the imagery adopted by the Madres in their speeches, women were perceived to be life-givers and guardians of life, taking especial care of their biological children with whom a direct and nurturing bond was assumed to exist. What is striking about this discourse is not its content, which bears resemblance to women's movements elsewhere in Latin America and in the world (e.g. Schirmer 1989), but the way in which place and politics became intertwined. In the elaboration of the Madres as a political movement, an elision between place and maternal identity was founded and maintained: Madres and Plaza became synonymous, and the socio-political meanings generated by women's position in the Plaza circulated around notions of mothering, life, and essentialist categories of kinship and peace. In the restriction of civilian public space, the Madres extended the domestic female sphere of influence; they did not transform one or the other in any lasting way.

This argument can be further illustrated in relation to the issue of gender violence. By extending the domestic dynamics of mothering duties into the restricted public space, the Madres did not generally voice any prioritizing of gender issues such as divisions of domestic labour, or male violence against women. In other words, the community of resistance developed in relation to the violation of familial-state relations, not in relation to women's intra-familial relations. The women's resistance movement, while able to question the legitimacy of the military regime, was, by the same dynamic, unable to create resistance to prevailing gender roles. When the Madres claimed the Plaza during their rounds, they pointed more to the illegitimacy of military force rather than to the illegitimacy of violent force in society, including men's violence against wives. The 'boots, slaps and the kicks' which prompted the Madres' action, mentioned by Hebe de Bonafini, were attacks from soldiers, *not* from husbands and partners. Domestic violence was not part of community discourse among the Madres although, as mentioned above, other aspects of their domestic lives were converted into public mobilizing symbols of opposition to the regime.

Certainly, in the Madres' actions there were changes in personal uses of spaces. To carry out the appropriation of the Plaza, women abandoned the domestic roles which had defined their activity spaces and political identities previously. Women's work in cooking, cleaning and shopping lost priority in relation to the self-assumed task of finding children. In the words of one Madre, 'from washing, ironing and cooking we went out into the street to fight for the lives of our children' (Feijoó and Gogna 1987: 151). In other words, women were prioritizing certain aspects of their identities (motherhood over domestic work) in voicing concerns and setting agendas for action.

In articulating priorities and demands within the context of motherhood, the Madres reproduced discourses of femininity and female activities which resonated through Argentine history. María del Carmen Feijoó (1989: 88) argues that the Madres reinforced gender divisions of labour and made female altruism more significant than demands for women's rights. She also perceives among the Madres a new style of sacrificing *marianismo*, imitating a mythical Virgin Mary and trapping women into an essentialist association with 'Life' (Stevens 1973; McCormack and Strathern 1980). Women in the Madres de Plaza de Mayo continued in the majority to perceive themselves as mothers and wives, rather than as fighters for female rights, thereby reinscribing themselves directly into the Catholic nationalism which informed much of Argentine history.

Nevertheless, some Madres' expressions of opposition to authoritarian interpersonal behaviour in the language of human rights ran parallel to the concerns voiced by Argentinian and Latin American feminists throughout the 1970s about male violence against women, as well as issues such as forced conscription and fathers' rights over children. As argued by the Chilean feminist Julieta Kirkwood, 'the concrete daily experience of women in Latin America *is* authoritarianism' (Kirkwood 1988: 19). In turn, feminist politics developed in relation to the various ramifications for women of this social authoritarianism, in some cases making linkages with the Madres. In 1984, feminists and Madres forged common cause over such issues as: the commemoration of International Women's Day; calls for the end to discrimination against women; pension rights for women working in the home; the creation of a Ministry of Women; and the return of children to their families (Chester 1986: 17).

The placing of domestic violence and authoritarianism on the agenda by feminists and human rights activists did not stop with the return to democratic government. Under Presidents Raul Alfonsín and Carlos Menem, legal measures to curb and prevent examples of 'domestic authoritarianism' were adopted. In 1991, the Senate in Buenos Aires province approved a law expelling from the home men guilty of beating women, including wives, partners, mothers and grandmothers, children and other relatives (*Mujer/Fempress* no. 117, July 1991). A survey found that 65 per cent of Argentinian women had been beaten by a man at least once, and over three-quarters of cases were found within marriage. In cases of persistent domestic violence against women, women's rights to defend themselves have been supported in the courts: in two separate incidents where women killed violent husbands, they were released (ibid.).

Negotiations over gender-specific demands under democratic rule in Latin America raise questions about the degree of transformation in political spaces and collective identities for women under different regimes. Under the military government of 1976–83, women devised their own methods of doing politics, what Feijoó calls 'hacer política desde las mujeres', in which voting, 'citizenship' and standard political discourse were put aside in favour of mobilization around mothering. While often seen to be a natural identity for the Madres, in an analytic sense, 'mothering' cannot be seen as a purely essentialist concept, as Liz Bondi (see Chapter 5) and Diane Fuss (1989) remind us. Rather, the social construction of the Madres' identity was through a selective discursive formation around notions of a 'naturalistic', Catholic, self-sacrificing motherhood in which ideologies about 'basic' family commitment were echoed by the military junta.

After the return to democracy, the Madres continued their political activities, calling for the punishment of army officers responsible for deaths and disappearances. Their opposition to the law granting immunity from prosecution for many officers is well known. The Madres' persistent remembering of the destruction of their families and their knowledge about kin challenged the 'sanitization' of recent history which emerged at this time, in the 'official version'.[16] Inserting themselves as subjects of history, the Madres claimed a role as witnesses and recorders of events which were denied officially and, in many cases, socially. The Mothers' careful recording of every disappearance, Nancy Sternbach terms an 'invasion of the space occupied by official history' (Sternbach 1991: 94).

In conclusion then, under military rule in Argentina, women of varied class and family backgrounds formed a coherent active group of resistance to that regime. Basing their struggles on a primary identification of motherhood, in which a universal maternal identity was postulated to unite women, they literally and symbolically transposed private/personal issues and identities into a public/political space. Taking symbols from a historically feminized domestic activity space into a traditionally male sphere entailed the overturning of meanings associated with those sites. In other words, vulnerable women became the strongest, most effective political actors at a time when classic civilian politics was 'underground'. Women were female in their self-representations and expectations, yet were not feminine sufferers and silenced, as the regime had expected. They were demonstrably political, although their mistrust of the term 'politics' was notorious.

However, this view of the Madres and their movement only takes us so far. While there was a transposition of domestic, childcaring roles into the streets and squares of Argentina, the identity which motivated and sustained these women was one of domesticity, female parenting roles and Catholic notions of duty and suffering. As female subjects of the Argentinian nation, with all the specific historical and political relations which that connotes, the Madres reinscribed their feminine identities in different places and with different activities during the military regime, but they did not fundamentally change their subjectivity. In other words, the move from 'private' to 'public' represented a shift in activity spaces, but did not transform their identities in relation to gender, power relations, nationalism or violence. The changes in identity experienced by most women of the Madres group were in the contingent relations between themselves and the regime, rather than in a fundamental restructuring of this relationship, through challenging notions of citizenship, power and gender relations. (These issues were much more systematically addressed by feminists.)[17]

Temporarily, women left pots and pans to look for children, as the military Catholic regime had exhorted them to. Metaphors of public and private, so embedded in Argentine cultural history, remained unchallenged by the Madres, whose own language picked up and reproduced such discourses. In this sense then, the underlying relationship between gender and politics remained more stable than is suggested by the literature on women's role in influencing regime transitions. Given the persistence of metaphors of place, gender and politics which circulated during the military regime in Argentina (despite the shifts in geographical relations between home and street during the same years), it would appear that the discursive, metaphorical relations articulated by Argentine nationalism have a great hold on political-personal identities. Michael Taussig suggests that Argentinian women's essentializing role as 'the nation' was re-articulated by the Madres, in what he terms a 'refunctioning of assumed essences' (Taussig 1990: 220). I would argue that the crucial point here is that the female essences (as subordinate citizens, as 'tradition', etc.) *remained*: nationalism was still fundamentally an essentialized relational identity of people with place, an identity which was *not* 'refunctioned' by the Madres.[18] Although this identity was reconstituted for some women (through their subsequent involvement in feminism), feminine identities at a national political level remained remarkably unchanged.

In this sense then, the emergence of an identity politics around being Mothers of the disappeared was contingent upon a particular configuration of sites (police stations, clandestine detention centres, homes and cafés) and power relations which gave rise to political mobilization in and around these places. As the political situation changed, the power relations and symbolisms around these places changed, although the Madres retained activist practices attuned to the military era. Already in the late 1980s, the Madres de Plaza de Mayo were accused by other women of inappropriate political methods (see Feijoó 1989: 73). Although the collective identity of the Madres provided an effective community of resistance to the military junta, the progressive transformative potential of such an identity revealed its limitations in the return to democracy and the opening of wider political debates, not least due to an underlying nationalism which constructed an homologous identity for women and the nation.

NOTES

1 It is estimated that at least 15,000 to 25,000 people died in Argentina in its 'dirty war' against civilians, between 1976 and 1983. Some 8,900 disappearances were officially established, and some 20,000 people were arrested. Two million people left the country during the military government.

2 For more details on the relationship between gender relations and regime change see Alvarez (1990), Jaquette (1989), Bourque (1989), and Radcliffe and Westwood (1993).

3 Among the Latin American literature, this argument can be found in Oria (1987), Diago (1988) and Calvera (1990).

4 De Lauretis (1987); also on the importance of context see Alvarez (1990) and Bourque (1989).
5 Moreover, in Argentina the Catholic left associated with liberation theology and similar movements represented only a small minority.
6 Perón's widow, María Estela Perón, then took charge of government until the military coup in early 1976.
7 Contempt would originally have been associated with racial 'difference', as gauchos partnered indigenous women (see Calvert and Calvert 1989: 141).
8 Women in Argentina were however active in first-wave feminism, organizing (among other things) journals for feminist ideas in the 1850s, and the First Female Congress in 1910 (Alvarez 1986).
9 Chile followed quickly behind Argentina, with Peru and Mexico granting female franchise in the mid-1950s and Colombia in 1957 (Jaquette 1989: 3).
10 Julieta Kirkwood sees this conservatism as culturally and politically 'induced', and attributes it to the inevitable result of an authoritarian patriarchal gender politics (Kirkwood 1988: 21).
11 My information on the organization of the Madres de Plaza de Mayo comes largely from Feijoó and Gogna (1987; 1990), Diago (1988), Oria (1987), Calvera (1990) and Feijoó (1989).
12 The Abuelas group developed in relation to a different geography of disappearance to the Madres, as grandmothers encountered each other in the juvenile courts and various children's homes (*LADOC* 1984: 29; also Oria 1987).
13 'We, the Madres, are not a feminist movement, because we are fighting for our children' (quoted in Calvera 1990: 105).
14 'Faith in God remained intact still. In that the lacerating pain took refuge' (Madres de Plaza de Mayo 1987: 8).
15 Several parallels can be seen here with the Brazilian case where women undertook mobilization and contestations of political meanings under the military, all the time denying that what they were doing was 'politics', as that was a male activity, involving corruption, disloyalty and so on, which they knew nothing about (Pires de Rio 1990; also Chaney 1979; Radcliffe and Westwood 1993; Radcliffe 1990).
16 This in turn was the title of an important film made to address precisely these issues.
17 The annual National Meetings of Women provide a space for the discussion of these issues, by thousands of women (6,000 in the 1991 meeting). Interestingly, in the context of this book, the women's movement in Argentina makes regular use of spatial metaphors in its publications and pronouncements. For example, the National Meetings go to a different region of the country each year, in order to increase their 'informative space', while the Buenos Aires provincial Women's Council is seen as 'an institutional space dedicated to women' (*Mujer/Fempress* no. 117 (1991); cf. Vargas 1991 on Peru).
18 In this light, a point which might be worth pursuing is the relationship between nationalism (which could be defined as a long-term largely metaphorical relationship with a given unchanging large geographically defined space) and female/woman (whose symbolic-metaphorical function in geography has been to represent stability and place, as opposed to time see Massey, Chapter 8).

REFERENCES

Alvarez, S. (1986) 'The politics of gender in Latin America: comparative perspectives on women in the Brazilian transition to democracy', unpublished PhD thesis, Yale University.

—— (1990) *Engendering Democracy in Brazil: Women's Movements in Transition Politics*, Princeton: Princeton University Press.

Bonafini, H. de (1985) *Historias de vida: Hebe de Bonafini*, ed. M. Sánchez, Buenos Aires: Fraterna.

Bondi, L. (1992) 'Locating identity politics', in session 'Communities of resistance: geography and a new politics of identity', paper given at the Institute of British Geographers Conference, University of Swansea, January.

Bourque, S. (1989) 'Gender and the state: perspectives from Latin America', in S. Charlton, J. Everett and K. Staudt (eds) *Women, the State and Development*, Albany: State University of New York.

Calvera, L. (1990) *Mujeres y feminismo en la Argentina*, Buenos Aires: Grupo Editorial Latinoamericano.

Calvert, S. and Calvert, P. (1989) *Argentina: Political Culture and Political Instability*, London: Macmillan.

Carlson, M. (1988) *¡Feminismo! The Women's Movement in Argentina from its Beginnings to Eva Perón*, Chicago: Academy Chicago Publishers.

Chaney, E. (1979) *Supermadre: Women in Politics in Latin America*, Austin: University of Texas Press.

Chester, S. (1986) 'The women's movement in Argentina: balance and strategies', in ISIS International *The Latin American Women's Movement: Reflections and Actions*, Santiago: ISIS.

Cubbitt, T. (1988) *Latin American Society*, London: Longman Educational.

de Lauretis, T. (1987) *Technologies of Gender*, London: Macmillan.

Diago, A. (1988) *Hebe: memoria y esperanza*, Buenos Aires: Ediciones Dialéctica.

Feijoó, M. C. (1988) 'Mujer y política en América Latina: el estado del arte', in *Mujeres latinoamericanas: diez ensayos y una historia colectiva*, Lima: Flora Tristan Centro de la mujer peruana.

—— (1989) 'The challenge of constructing civilian peace: women and democracy in Argentina', in J. Jaquette (ed.) *The Women's Movement in Latin America: Feminism and the Transition to Democracy*, London: Unwin Hyman.

Feijoó, M. C. and Gogna, M. (1987) 'Las mujeres en transición a la democracia', in E. Jelin (ed.) *Ciudadanía e identidad: las mujeres en los movimientos sociales latinoamericanos*, Geneva: UNRISD.

—— —— (1990) 'Women in transition to democracy', in E. Jelin (ed.) *Women and Social Change in Latin America*, London: UNRISD/Zed.

Fletcher, L. (1992) 'La forma genéricamente específica de tortura (la violación) durante la última dictadura militar en Argentina' *Fem* no. 109, March (Mexico).

Fuss, D. (1989) *Essentially Speaking: Feminism, Nature and Difference*, London: Routledge.

ISIS International (1986) *The Latin American Women's Movement: Reflections and Actions*, Santiago: ISIS International.

Jelin, E. (ed.) (1989) *Ciudadanía e identidad: las mujeres en los movimientos sociales latinoamericanos*, Geneva: UNRISD.

Jelin, E. (1991) 'Social relations of consumption: the urban popular household', in E. Jelin (ed.) *Family, Household and Gender Relations in Latin America*, London: Kegan Paul International/UNESCO.

Kirkwood, J. (1988) 'Feministas y políticas', in *Mujeres latinoamericanas: diez ensayos y una historia colectiva*, Lima: Flora Tristan Centro de la mujer peruana.

LADOC (1984) 'Grandmothers of the Plaza de Mayo: sorrow and hope', *LADOC* XV: 29–37.

McCormack, C. and Strathern, M. (eds) (1980) *Nature, Culture and Gender*, Cambridge: Cambridge University Press.

Madres de Plaza de Mayo (1987) *Nuestros hijos*, Buenos Aires: Editorial Contrapunto.

Medina, C. (1989) 'Women's rights as human rights: Latin American countries and the Organization of American States (OAS)', mimeo.
O'Donnell, G. (1973) *Modernization and Bureaucratic Authoritarianism: Studies in South American Politics*, Berkeley: University of California Press.
Oria, P. (1987) *De la casa a la plaza*, Buenos Aires: Editorial Nueva América.
Pires de Rio, T. (1990) 'Women, daily life and politics', in E. Jelin (ed.) *Women and Social Change in Latin America*, London: UNRISD/Zed.
Radcliffe, S. (1990) 'Multiple identities and negotiation over gender: female peasant union leaders in Peru', *Bulletin of Latin American Research* 9(2): 229–47.
Radcliffe, S. and Westwood, S. (1993) *Viva! Women and Popular Protest in Latin America*, London: Routledge.
Rowe, W. and Schelling, V. (1991) *Memory and Modernity: Popular Culture in Latin America*, London: Verso.
Sarmiento, D. (1845) *Facundo*, Buenos Aires.
Schirmer, J. (1989) ' "Those who die for life cannot be called dead": Women and human rights protest in Latin America', *Feminist Review* 32: 3–29, Summer.
Sternbach, N. (1991) 'Re-membering the dead: women's testimonial literature', *Latin American Perspectives* 18(3): 91–102.
Stevens, E. P. (1973) 'Marianismo: the other face of machismo in Latin America', in A. Pescatello (ed.) *Male and Female in Latin America*, Pittsburgh: University of Pittsburgh Press.
Taussig. M. (1990) 'Violence and resistance in the Americas: the legacy of Conquest', *Journal of Historical Sociology* 3(3): 209–24.
Vargas, G. (1991) 'The women's movement in Peru: streams, spaces and knots', *European Review of Latin American and Caribbean Studies* 50: 7–50, June.
Westwood, S. (1989) 'Women, nation and the state', paper given at the Institute of Latin American Studies, London, at the conference 'Women in Popular Protest in Contemporary Latin America'.

7

READING *ROSEHILL*

Community, identity and inner-city Derby

George Revill

> Mohammed's House, a small villa in a terrace of villas, is set nearly half-way down the side of a hill in a wide sloping street. At the top stands a telephone box, and the big back gate of the vicarage. At the bottom of the street is a small post office, across the road from The Duke of Cambridge, the public house much patronised by football fans before and after the match. It's only a few years since I have lived there, but that time hangs now in my memory like a golden globe.
>
> On the first day, I woke up on the big high bed, with my possessions piled up at the bottom, afloat on a sea of old yellow and black lino . . .
>
> By standing on the bed I had a great view – chimneys and chimneys, a bit of the church, and a back garden where swarms of little Pakistani boys in Fair Isle pullovers climed about all over the outside of their house, like the old woman who lived in a shoe's children did . . .
>
> It is autumn and very warm. I stroll about the streets and the Arboretum, watching the yellow leaves toss about.
>
> At nights it is noisy; the wind howls round the rooftops and you notice it more in an attic; the doors shudder and rattle and the light bulb swings in the wind. Dogs rave at one another at periodic intervals throughout the night, and there is a sound I can't identify which resembles an old-fashioned steam train or an owl with a megaphone. There is a canine dawn chorus, and after that comes the church bell on Sundays. On Saturday afternoons in the autumn and winter comes the roar from the football ground, every time the local team score, and if I stand on the bottom of the bed I can see the floodlights, high into the sky.
>
> (Lake 1989: 43–4)

In this manner Carol Lake describes her new life in the Rosehill district of Derby. Her book of short stories, *Rosehill: Portraits from a Midlands City*, was published in 1989. It was written as a form of diary and relates to the period 1985–6 when she lived in a rented room in this inner-city district. It was well received, winning the *Guardian* prize for fiction in 1989. It had quite extensive reviews in *The London Review of Books* and in *New Statesman and Society*, both of which published individual stories. This chapter examines the way in which Carol Lake employs both the place in which she lives and the people who live there as part of the development and articulation of her projected personal identity. The chapter is divided into two sections: the second part presents a reading of the Rosehill stories and considers the relationship between community, identity and story-telling: the first part presents a commentary on that reading and the circumstances in which it was produced.[1]

LOCATING ROSEHILL

I first read *Rosehill* as a postgraduate student when I was undertaking research on this locality (Rosehill in the nineteenth century). At this time it was a high-status residential district for railwaymen, who migrated from all parts of the country to work here at the headquarters of the Midland Railway Company. The Rosehill that Carol Lake portrayed was rewardingly familiar to me. To the academic, it was a locality constituted of migrants bringing with them a multiplicity of cultures, an urban fabric which retained the symbolic foci which had articulated its social life in the nineteenth century. To the student, it projected the realities of bedsit land and the traumas of life on state 'benefits' in Thatcher's Britain of the 1980s. But what most impressed me was Lake's ability to articulate the conditions in which people live in this district without making them the objects of social investigation; to inscribe personal identity in a most public forum without rendering either herself or the community she describes transparent or readily appropriable by the gaze of authority. This appeared to be founded not on an overt theoretical correctness but on a power of geographical imagination, to hide the mundane in the exotic and to secrete what has personal value in a quotation web of routine. To a geographer struggling to write something of personal significance by attempting to apply a desk full of exotic theory to an archive full of very humdrum data, this was indeed something to be admired.

In the nineteenth century, Rosehill was a prosperous Derby suburb built around the Arboretum and the Infirmary. Clustering on the top of a ridge around these institutions, the district formed a highly respectable residential area of detached villas and was soon surrounded by better artisan-type dwellings running down the slopes away from the Aboretum. Caught between an aero-engine factory and the railway engineering workshops, the district was severely bombed during the Second World War. Today, over 54 per cent of the population come from a New Commonwealth background. Though there are a large number of Muslims, the substantial majority are Sikhs from villages in the East Punjab. A Derbyshire Social Services report of 1975 followed by the Census in 1981 gave a statistical picture of urban deprivation seized upon by the local press. Children in care numbered 15 per cent of the county total, whilst the district accounted for 30 per cent of Derby's psychiatric patients. Levels of unemployment have ranged above the 35 per cent mark throughout the 1980s. Many of the derelict sites and slum clearance areas remained from the postwar period into the late 1970s, and only in the 1980s have a number of rather piecemeal urban housing schemes brought better quality housing. Many of the older institutional buildings have been turned to new use, becoming multicultural centres, Sikh temples and mosques.

Carol Lake appears to slip into this world quite easily, from a single-parent family, having spent some time in a Derby mental hospital in her young adult life, a period alone living in an inner area of neighbouring Nottingham, and on the dole for ten years, she returns to the area where she was brought up as a child. However, this book is not simply a rediscovery of roots or just a means of expunging the past by reliving it. Reviews of the book latch on to its role as factual reportage: 'it is really real', according to one reviewer – a documentary in 13 chapters 'which can be trawled for the sort of information collected by feature writers' (Clapp 1989). Representing the experience of everyday life, the stories frequently adopt colloquial and slang language tinged with childishness. The stories frequently lack a sense of direction, telling of mundane and routine activity, fickle changes of mind and mood, and fortuitous coincidences which often have no clearly positive outcome or even any conclusion. Any set of short stories which manages in so few words to deal with unemployment, drug-taking, divorce, destitution, prostitution, racial violence and female circumcision is perhaps bound to be classified as a documentary. Yet, if it is a documentary, it is a far from simple one. 'The Day of Judgement', for example, relates the story of God arriving in Rosehill during a latter-day flood, getting out and going to the pub while the Ark sails away with its full complement of saved souls.

This chapter focuses on the concept of community, and as part of a volume on the topic of place and the politics of identity, this is justified for several reasons. Frequently, community has been championed as a source of identity, of moral and social stability, of shared meaning and mutual co-operation. Yet, the concept of community has forceful negative connotations.

First, community can be thought of as a threat to identity as it has been articulated by conservative and reactionary thought. From the origins of the welfare state to 'care in the community', the concept has been a morally charged instrument of authority justifying state intervention in everyday life (Plant *et al.* 1980; Nisbit 1962; 1967). Romantic and nostalgic thinking has often resorted to the invocation of a lost stable social hierarchy of community in order to justify socially repressive policies (Wiener 1981; Colls and Dodd 1986).

Second, because community is so often viewed as something static and parochial, it poses limits on identity controlled by tradition and passively accepted local culture. Even as the source of class-based and anti-establishment strength, community can be viewed as subordinating the individual to the whole subordinating differences for the sake of communal solidarity (Laclau and Mouffe 1985). Community can therefore be viewed as a vehicle for the reproduction and perpetuation of 'traditional' gendered social roles; 'the nuclear family'; the subsidiary role of women in a male-dominated society.

Studies of community are strongly associated with studies of locality. Strong arguments exist which claim that the academic study of localities (and by implication communities) reproduce the very romantic, parochial and repressive version of society which critical enquiry should seek to break down. However, it is argued here that, for good or ill, the idea of community does have a part to play in the way people think about themselves, in the construction of subjectivity, and in the production of personal identity. Many of the most articulate opponents of local studies recognize this whilst objecting to the reactionary implications of community (Harvey 1989b: 231). However, working with the concept of community does not necessarily imply romanticism, reaction or stability. Massey's defence of locality studies has relevance for the notion of community. She argues that places have multiple meanings and that it is important to think of localities (and arguably communities) in terms of fluidity, contradiction and conflict (Massey 1991: 275).

The criticism of locality studies that associates them with the static and bounded can be traced to a version of place in which it has become equated with space as a universal and transhistorical construct based on an interpretation of the Heideggerian notion of Being which implies fixity and stasis. Locality has thus been characterized as 'the foundation for collective memory, for all those manifestations of place-bound nostalgias that infect our images of the country and the city, of region, milieu, and locality' (Harvey 1989a: 218). Interpreting the relationship between a stable identity and a static and bounded place as one in which the former follows mechanically from the latter as a 'seamless coherence', bestows a sense of causal necessity on the relationship. The unexamined acceptance of this forecloses the important questions of identity formation by which places are socially constructed. Far from intravenously infecting us with nostalgia, place is an active process of fixation: this operates through social action at a variety of scales of which a nationally embedded culture of nostalgia is but one.

The value of community as a concept in this context is that it throws into prominence the tensions between senses of belonging which form ties between individuals and groups and between peoples and places. It is not that it enables us to identify a stable or even dominant set of social and cultural characteristics by which a particular place or group of people might be identified. Rather, community focuses interest on the processes that create a sense of stability from a contested terrain in which versions of place and notions of identity are supported by different groups and individuals with varying powers to articulate their positions (Jackson 1991a). This approach is adopted by Wright (1985) in his writing on inner-city London. He is concerned with the politically charged role a sense of belonging plays in contemporary English society and with the means by which senses of belonging develop from differing engagements with past and present, from local, national and international economy, from society and culture.[2] I find the Rosehill stories interesting because they do not simply romanticize community as a caring organic society of shared consciousness, though this aspect of the concept is acknowledged. They recognize separate ways of both living and imagining a specific locality which are fluid, permeable and conflictual. Most importantly, they recognize the role of the distant in the construction of the local, drawing out the strands of consciousness represented by story-telling, and demonstrate the role of community as a way of mediating these strands. In doing this, they explore the role of story-telling in the process of fixing a sense of identity in place.

Patrick Wright adopts an approach to Being which views Being as a more active concept than even the defenders of locality studies will admit (1985: 5–14). In so doing, he follows Heller's materialist philosophy of 'everyday life'. She recognizes that through a process of 'objectification' and 'integration', individuals constantly reproduce themselves as social beings (Heller 1984; Revill 1991: 73–5).[3] She shares with Heidegger the idea that human beings are inescapably located and always begin to make sense of themselves from a position of already-being-in-the-world. Concerned to break the Cartesian dichotomy between mind and matter, she argues for the impossibility of detachment by recognizing the empirical necessity of involvement in the world if an individual is to function as an independent being. Her reading of Heidegger is here quite close to that of Paul Ricoeur (1981: 185–90), who claims that understanding begins with 'inhabiting-the-world'. In what he calls a dialectic of distanciation and appropriation, self-understanding (and by implication self-identity) is continuously made and remade through a process of assimilating the inhabited world and projecting conceptions of what is known back into the world. The subject 'plays' with meaning in order to gain greater understanding. Story-telling is a form of such 'play'. Wright's concern is with the processes by which senses of belonging and notions of identity develop from an engagement with the world inseparably both material and imaginal. For both Heller and Wright, story-telling is an important means by which we make sense of the world, appropriating our environment and finding a location within it. The Rosehill stories form both the *active* autobiography of the author (in the sense that she consciously relates her own history) and her *passive* autobiography (in the sense that their diary format documents two years' work). Given their autobiographical content, these stories are important because they describe such an engagement with the world. They record one woman's struggle for identity (a struggle it is easy to lose in the word play): both a physical identity for someone surviving on dead-end jobs and 'social security', and an intellectual identity as an author finds a voice.

If one adopts the conceptual framework charted above, then because Being always starts for the individual from a specific empirical situation in-the-world, the formation of individual knowledge is always contextual and contingent on the individual's location in the world. Therefore, no amount of transcendental regression can take a subject back to the position where thought has sufficient purity to gain intersubjective authority. This is because such a degree of sharedness can only be claimed from an initial position of passivity, where in the first instance individuals receive sensory information as inert receptors. There is, therefore, an inescapable gulf between individuals in which others are ultimately unknowable. This implies that if it is impossible to claim one can fully understand the motivation of another on the grounds of shared knowledge, it is not possible to include the words or text of another on the grounds that they express the intentions of another. Where someone else's words inhabit a text, they serve only the purposes of the author and a range of unintended purposes which escape intention. Extensive quotation or a polyvocal writing style, frequently championed in anthropology for example, cannot fully and consciously express the intentions of a person other than an author (Marcus and Fischer 1986; Clifford 1988). At the moment when another voice appears to speak with the greatest clarity and highest degree of naturalism in a text, it is arguable that the words most ably express only the literary dexterity of the author. Support of this position comes from writers arguing from a variety of perspectives: based simply on the power of the author as final arbiter of the textual product (Marcus and Fischer 1986: 68; Clifford 1988: 68); based on the inability of a finished text to capture the open-ended creative character of the active flux of dialogical communication (Ricoeur 1971); and based on ethico-political grounds. Many writers recognize that the procedures of Western intellectual endeavour inevitably burden the source of their enquiry with a weight of alien conceptual material which disables even the most articulate and vociferous other from speaking through the text (Lyotard 1988; Spivak 1988; Young 1990).

Yet there is equal argument which casts this relationship of separation in a positive light. Ricoeur (1981: 131–44) argues that distanciation is necessary to the establishment of a critical distance fundamental to rational academic enquiry (though this perspective is founded on the now much-criticized possibility of intellectual detachment). Lyotard (1988) believes that through the philosophy of the Differend founded on a heteroglotia of phrases juxtaposed in their dissimilarity, voice can be given to minority discourses, preserving rather than suppressing differences. Bakhtin (1981: 293) has suggested that language is inescapably intersubjective as it 'lies on the borderline between oneself and the other. The word in language is half someone else's' and is 'completely shot through with intentions and accents'. Both these positions can conflate the formal traces of the dialogue of others with the preservation of the intentions bound up in lost locutionary acts. hooks (1991: 146) views the marginalized position of otherness as offering a vital opportunity, a position of privilege from which one can gain authority to articulate the position of the un/misrepresented. Yet, echoing Said (1978) and Spivak (1987; 1988), she poses the moral dilemma resultant from the activity of translating and communicating the position of the oppressed, even assuming the authority to speak. When one needs the oppressor's language to communicate, to articulate a position is to reproduce the language of domination and, in consequence, the fact of domination itself.

This is an important issue for this chapter in terms of both Carol Lake's relationship with the characters in her stories and my own relationship with Lake's text. Carol Lake is adamant that she is in control of her own writing, and as a piece of self-consciously autobiographical fiction this is perhaps not surprising. However, her writing has not been universally accepted by those who see themselves depicted in it. Her dual position as an insider relating her own thoughts and as an outsider consciously researching a book puts her in a privileged position where 'fiction' defends 'fact', and where apparently 'factual' statements can be justified as the product and property of an individual imagination. As a reader of *Rosehill* and also as its writer, my position is also complex. It is argued that stories, even written narratives, are constructed as a contested terrain negotiated between the teller and the told, between the writer and audience, in which the story-teller partly adapts the text for the reception of an audience and the audience actively works the meaning of the tale in terms of their conscious and unconscious motivation for reading and their expectations of the text (Barthes 1977: 142–8; Rimmon-Kenan 1983; Chambers 1984; Moi 1985: 24; Maclean 1988: 13–47).

This piece of writing is necessarily located by my own motivations for reading Carol Lake's *Rosehill* and my own attempts to address a particular audience. Conscious reasons for choosing to write about Rosehill include familiarity with the district, an identification with the circumstances under which the author lived, and also an admission that the stories 'flatter' a particular set of ideas which were part of my own intellectual 'development'. These originate in my own engagement with Rosehill as a postgraduate student. In a curious way, I identified a course in which our very different Rosehills functioned homologically in the creation of our respective self-identities. Rosehill functions this way at this very moment, allowing me to intervene in the construction of myself as an academic. One theme of this chapter is the role of story-telling in overcoming the gulf between self and other, past and present, near and distant. This chapter can be viewed quite easily as such a story. It links me to the historical and geographical source of my academic career and it justifies a particular approach to the subject matter: it is an attempt to give meaning and purpose to an inevitably large part of my life. In so doing, the writing reaches out to 'grasp' the experience of another. This is in the sense that a progressively organized, 'reasoned' argument is an attempt to gain understanding and therefore also an attempt at mastery and control. This writing therefore commits violence upon that which it purports to celebrate, defining Carol Lake's text in alien concepts and structuring it in linear argument. In its unifying, homogenizing drive for understanding, an unrequited empathy for the subject is translated into an unwanted patronage.

It is important therefore to recognize this 'desire' as a danger as well as an inescapable consequence (Hearn 1987; Jardine and Smith 1987; Belsey and Moore 1989; Jackson 1991b). It is an ethical problem founded on an unescapable void.

The philosopher and Talmudic scholar Emmanuel Levinas has endeavoured to set out the basis for an ethical relationship between self and other which does not aim to eradicate difference (Levinas 1981; Bernasconi and Wood 1988; Bernasconi and Critchley 1991). He argues that there is a form of truth that is totally alien to me, that I do not discover within myself, and that requires me to leave the realms of the known if I wish to address it. This other is not a threat to be reduced or an object that I give myself to know in my capacity as knowing subject, but that which constitutes me as an ethical being: in my originary encounter I discover my responsibility for the existence of this other, a responsibility that will lie at the root of all my subsequent ethical decisions. It is my capacity to communicate with others that has been opened up by the relation, and in this encounter reason is chastized. It is not likely to seek hegemonic control, for were it to do so, it would have to do violence to my self as the self that is in this relation of response-ability to the other (Godzich 1986: xvi). Levinas sees the ethical moment as that which remains after that which can be accounted for ontologically has been taken away. This suggests that it is founded on a phenomenological intelligibility based on a transcendental intersubjectivity that precedes language. His project is criticized by Derrida. Derrida says that because the tools used to make such a critique are the concepts and categories of the Western intellectual tradition, and are necessarily embodied in language, the originary, transcendental, ethical moment remains inescapably distant (Norris 1987: 233; Derrida 1991).

I am unable to accept the transcendental foundations of Levinas's ethical relationship; and I am unable to justify on the grounds of my own marginality my right to speak about a woman writing out her experience of a multicultural inner city. Yet Levinas voices a relationship of absence which resonates strongly with the location I find myself. The biological/physical and political/theoretical void of this differences is expressed by Heath (in Jardine and Smith 1987: 1) as a barrier to communication, yet this space seems also to be a powerful motivational force. I still wish to base my response to this book on some ethical grounds. The origin of that desire could be located in my socialization as a male in the fascination with the exotic which has been historically important to the construction of male hegemony. If desire is central to human motivation in the contemporary world, then it must be recognized that desire is a creative force, with violent and destructive possibilities (Sennet 1977; Deleuze and Guattari 1983, 1988; Lasch 1985). At its centre lie the anticipation, expectation and speculation generated by a non-presence. Story-telling not only *bridges* space in the sense that it links disparate elements by imaginative threads of reasoning, but it also *creates* space in the sense that by the act of signification, inscribing experience in words, new bifurcations are produced (de Certeau 1984: 122–30). These are the source of new voids to be crossed and desire is both satisfied and reinvigorated. If the origin of my desire to read *Rosehill* is based on my male socialization (a potentially destructive force), then perhaps my desire for an ethical response (a potentially benign force) is also based on the perpetual deferment of gratification dependent on the endlessly deferred product of the signifying process. This desire motivates the acts of reading and writing and the space opened by them. If the structuring act of story-telling produces spaces to be crossed, and if I am defined by the volumes and boundaries produced by that structuring – by that which is other than I am – then that 'form of truth which is totally alien to me' arises in the space opened by writing. The space that is intrinsic to the production of self and also the construction of otherness. My response-ability for the existence of the other in Levinasian terms can be found in language and in the creative/destructive powers of story-telling.

In terms of writing and reading, the openness of a relationship is a consequence of producing a text. If reading is a creative process, then a text is open to creativity whenever it is read, to a potentially infinite number of readings. If to be open, to give of oneself, is the Levinasian basis of ethics, then to recognize the openness of one's writing and to release it to the reader is of fundamental importance. If the production of language and the construction of texts is central to the process of self-definition, then to abandon what has been written to the desires of a readership is to place one's identity in the trust of the reader. This text is unavoidably the product of a creative reading of *Rosehill*, I take from Carol Lake's stories that which I think is important, that which tells my story and charts my path to Rosehill. Long quotations, like the one that heads this essay, give an illusion of polyvocality but also set the mood and tempo of reading I wish to project. A discussion such as this gives the illusion vulnerability but this is a calculated vulnerability: I endeavour to close certain readings of my text and enforce a reading of me through the text which I find agreeable. I am unable, not to say unwilling, to release control of my own text: it is argued above that in one sense this is theoretically impossible. However, the act of writing decrees that I have to give up my text. All I can do is admit the situatedness of any reading produced by my location and I hope that this writing, by forging a thread of narrative, has created a sense of anticipation, has opened a space, created the necessary void (for what Levinas might call admiration), such that you will desire to read *Rosehill* for yourself.

READING *ROSEHILL*

Reviewers of Carol Lake's *Rosehill* certainly believe that she is committed to the community in which she was brought up and to which she has returned to live. There is arguably much common ground between this set of stories and previous writing which focuses on community, particularly that on urban, industrial, working-class community. The fly-on-the-wall style echoes the scientific recording of an alien way of life found in mass observation, and in classics of the postwar period: *The Uses of Literacy* and *Coal is Our Life* (Hoggart 1957; Dennis *et al.* 1956). The detached factual documentation which paints a naturalistic picture but disguises and obscures this with false names, dates and places mirrors much work of the 1950s and 1960s: for example, that collected by Frankenberg (1966). Stress on the conversational as an arena for exercising the sharedness of community reflects the work of social anthropologists on rural Britain, while a focus on the micro-social, reflects that of recent studies of inner-city life, such as Sandra Wallman's *Eight London Households* (1984). The book has some of the 1960s starkness of Alan Sillitoe's (1988) short stories, and much of the aimless perambulation and open-ended pathos that infuses the films of Mike Leigh in their intimate exploration of English society. However, these stories do not have the onwards and upwards trajectory of escape and self-realization which infuses much male working-class literature, from Robert Tressell's *The Ragged Trousered Philanthropist* (1955) and Robert Robert's *The Classic Slum* (1971) to Braine's *Room at the Top* (1957) or Sillitoe's *Saturday Night and Sunday Morning* (1958). They do not strive to leave their 'place' neither are they satisfied with it. The stories are circumscribed by a world partly defined as a stereotypically female domain (launderette, child-minding, shopping), partly by a stereotypically urban rootless world of bedsits and job centres, and partly by the public institutions of urban community (pub, soccer ground, bingo, café, park, evangelical mission and mosque). They describe a circularity which one might attribute to a consciously gendered statement encapsulating non-male values; to an acceptance of a traditional female role where one accepts one's lot and makes the best of what one has; or to the economic and social deprivation of the poverty trap where a sub-proletariat, a reserve labour force, is maintained within the system by structures that prevent escape.

During the late 1960s and 1970s, academic sociologists seem to have dismissed the idea of community (Clarke 1973; Harper 1989). Margaret Stacey (1969), for instance, was unable to locate a satisfactory definition for community, amongst 94 definitions classified by place, social activity, social structure or communal sentiment. However, in spite of the intense disagreement, there are certain things in common amongst the sociological definitions of community. One can read into the locality that Carol Lake describes much that accords with a traditional conception of community and with its positive associations of a moral social order. I shall consider these under three headings:

1 informal social networks;
2 multi-dimensional experience of people;
3 a sense of history.

Informal social networks

Within traditional communities, information networks based on family and neighbours are believed to be of paramount importance in social control and communal censure and for sharing knowledge useful to overcoming the routine problems of everyday life, in making and earning a living, health care and home-making. Carol Lake's *Rosehill* is a world of backdoor gossip of chance encounters and casual meetings. Events from local riots, births, deaths, marriages and partings as well as the coming of the Ark are transmitted by this means.

National and international events are always articulated through local channels of communication, events half-heard on the radio or TV. Chernobyl, Handsworth, become local as they are mixed into conversations bound into the day-to-day problems of the community:

> Ahmed comes round. It's Saturday and he's spent all morning looking for a man who changes the colour of your shoes. He says there's a big Sikh procession in Rosehill, men with swords on floats.
>
> Like May Day, you mean?
>
> I don't know. They didn't look very pleased. They're still mad about Mrs. Gandhi.
>
> (Lake 1989: 9)

However, local communication networks do not always tell a story of communal innocence: the local phone box is used as a pick-up point for prostitutes and for the creation of casual relationships; girls sit beside it during the evenings waiting for people to dial the number and make a date.

Multi-dimensional experience of people

The experience of others in a variety of situations and in a multiplicity of roles and attitudes is believed to be an important quality of traditional communities where people come to be known in a variety of social contexts, as fellow workers, leisure-time friends, and the kin of an extended family system. Lake's characters are multi-faceted and often conventional expectations of social order are turned on their head. There are few black-and-white issues, few good or bad people: the 'Old Man' who lived at Mohammed's could be an object of pity but is a liar and lives in a world of pornography and toy cars; Mohammed the greedy landlord, defends and protects some of his white residents who are much less able to cope with the world than he is; Ahmed, the child hero and friend to Carol, is pompous and moralizing; the children in the street are, in turn, innocents gambolling in a wild garden and bitchy, thieving and street-wise.

A sense of history

The importance of historically situated social practices to the maintenance of community organization is both as a store of knowledge and as a justification of the continuance or repetition of practices. For Carol Lake, the traces of the past are always imminent in the present of Rosehill; sometimes this takes on a golden and nostalgic tinge creating a sense of communal certainty and permanence. She walks in the Arboretum and remembers when she and her friends used to be there in the 1960s as young and carefree adults. At the bonfire on wasteland, she remembers her childhood. However, sometimes the past becomes a spectre of great menace threatening to break any communal security. In her room a message written on the frame of an old mirror 'FLORENCE, Don't break the glass.' is fascinating and reassuring because it is an undecipherable link into the life of someone who used to occupy the room. But it is also menacing, because it echoes what she feels is her own fragile grasp on reality. On the street the sight of former fellow patients from the mental hospital are a constant and unwelcome threat to her identity.

Community, identity and story-telling

Community is a morally charged concept because it is about the obligations and expectations one has to those people one lives closest to and with whom one shares most in day-to-day life (Newby and Bell 1971: 218–49; Williams 1976: 75–6; Eyles 1985). The fact that we might think of community as an important means of influencing the social order is interesting precisely because it forges links between individual action and society; it links personal responsibility, commitment and identification with other people. In a so-called age of 'mass society' after the breakdown of the bonds of communality and shared purpose derived from religion and science, the morality that informs the individual in everyday life no longer has any foundations in certainty and we cease to view ourselves as bound within an 'organic' relationship with others and see ourselves as isolated islands of self-expression.[4] One may argue then that a claim for the end of community, just like a claim for its renewed potency, has implications which reach back to the core of what constitutes us as people. The social construction of self-identity has become a topic of some concern. To a great extent, this interest revolves around the problem of ontological certainty in a world of insecurity. This discussion of community comes at a time when both the foundations of morality and the construction of the self have come into focus (Kotre 1984; Lasch 1985; Wood and Zurchner 1988; Rose 1990; Giddens 1991; Baynes *et al.* 1987: 383–442; MacIntyre 1981). It comes at a time when critics like Edward Said warn against the seductive powers of religious fundamentalism to create communities of meaning and security in a world racked by crisis. It is also a cause of particular concern that a politics of the right has threatened our basic human rights in Britain with 'Care in the Community' and the 'Community Charge'.

Yet community still exists, if only because it is something people appear to want to believe – even if this is merely a post-industrial nostalgia for an industrial and pre-industrial past (Wright 1985; Hewison 1987). Though many people may admit to feeling positive about the notion of community, this is not to say that everyone means the same thing when they use the same words. Nor is it to claim that a shared sense of purpose, consent to action, means a consensual commitment to a common intersubjectively transparent set of social aspirations. Recent work by social anthropologists has addressed the tension between conformity and individuality within even so-called 'traditional' communities. The importance of community's symbolic rallying points are recognized precisely because they are polyvalent, providing an all-embracing concept which can contain the multiplicity of individual objectives and expectations (Cohen 1985: 21).

Sociologists and geographers may find community somewhat difficult to locate in modern society, where geographical and social boundaries are hard to draw on a collective basis. Yet if community as a morally charged concept is important to the construction of self-identity, then perhaps a rewarding way to think about it in the contemporary urban world is from the perspective of the individual rather than from that of the collective. A number of authors now point to the importance of biography to the construction of personal identity, the justification of previous action, and the continuity of the self into the future (Harré 1983: 213–14; Gergan and Davis 1985: 259–63; Giddens 1991: 47–55). As Charles Taylor (1989: 54) says, 'In order to have a sense of who we are, we have to have a notion of how we have become, and of where we are going'. In her book on depression, the psychologist Dorothy Rowe (1982: 112–24) stresses the part that autobiographical stories play in continued self-justification, through giving direction to our very existence. She views this in relation to those crises which potentially mark breaks in self-identity. Crises range from the unbridgeable gulf that physically separates us from our environment to the fractures that punctuate the trajectory of the self in the modern world: from birth, through separation from the mother, to marriage, retirement and death. For her, the stories we tell about ourselves are a means of transcending the fissures that separate past from future, the local from the distant. Such stories help to build a world of ontological security and continuity where the individual is situated in an ordered universe and feels capable to continue the act of living. This does not imply that the stories we tell about ourselves are complete, or that they are not mutually or internally contradictory. It does not mean that they are simply the product of a universal free will, or that they are cumulative in the sense that they add in to a permanent and fixed sense of self. Rather that they are strategically deployed by, and for the moment the product of, a specific cultural situation and the determinations suggested by this.

Community can be thought of as important to this sense of ontological security. Cohen (1985: 21), for example, interprets community always in relation to wider cultural and social processes 'consciousness of the world beyond is the catalyst for the recognition of one's own community as a discrete entity', we-ness is always defined in relation to otherness. Social and physical boundaries and the rituals that define them therefore become of paramount importance to the construction of community. Consequently community can be thought of as important to personal ontological security because it is about defining and ordering relationships between me and you, us and them. There would appear to be some relationship between the stories that define ourselves as people and symbolically meaningful activities which delimit community. Michael de Certeau's (1984) analysis of the spatiality of story-telling may help forge a link between the individual and the creation of community as a psychological, social and spatial entity. Drawing on a wide range of influences (including Marx, Freud, Merleau-Ponty and Derrida), he says, 'Stories traverse and organize places; they select and link places together making sentences and itineraries out of them.' 'They are spatial trajectories' founding and articulating spaces. He also stresses the importance of narrative to the articulation of everyday activities in a practical sense, where to name and describe is both an act of appropriation and an expression of dexterity, the exercise of know-how or commonsense knowledge. As de Certeau says: 'In this organization, the story plays a decisive role. It "describes", but "every description is more than a fixation," it is "a culturally creative act" ' (de Certeau 1984: 123). Stories create the world in which we live by defining its limits and boundaries: he calls these definitions 'frontiers'. Stories also link individual islands of activity and meaning together by passages of causal explanatory description and these passages he labels 'bridges'. This is not a secure world, it only lasts as long as the story is remembered and every time it is retold the world is created anew. But it is a certain world because it is based on the narrative process by which we describe the world to ourselves in our own terms to our own satisfaction, enabling us both to manipulate that world and to move around in it (de Certeau 1984: 122–30).

Ganguly's discussion of migrant identity, personal memory and the construction of selfhood works substantially within this framework. She is concerned with the way in which postcolonial migrants from India to the USA construct narratives of their premigratory past in the deployment of postmigratory identity. She says:

> My arguments centre on the proposition that recollections of the past serve as the active ideological terrain on which people represent themselves to themselves. The past acquires a more marked salience with subjects for whom categories of the present have been made unusually unstable or unpredictable, as a consequence of the displacement enforced by postcolonial and migrant circumstances. For my ethnographic informants, the present acquires its meaning only with reference to a disjointed and conflicted narrative of the past – in which references

to official narratives about colonization and a historical memory are tangled up with personal memories and private recollections of past experience.

(Ganguly 1992: 29–30)

These narratives of history not only work to establish a sense of community and shared identity amongst this group as a whole, but also serve as substantially personal justifications which present a strongly gendered reading of the past. Men disparage the past in order to construct themselves as self-made and migration as a positive experience of self-fulfilment. Women idealize the past because they refer to a time when they had clearly defined status and support systems within the patriarchal family.

It is in the context of community as a property of the individual and as important to the articulation of self-identity that I think we can come to terms with many aspects of Carol Lake's portrayal of Rosehill. One may suggest that the drive to establish ontological security would be of paramount importance to Carol Lake, given her history of mental illness and her return to home ground. However, her stories inhabit a realm at the margins of a material world and constantly test the boundaries of Western ideas on reality. Yet the places and people constructed in these stories are in a sense firmly located in Rosehill. It is through the notion of story-telling as a personal means of creating community as a physical and social entity that I want to continue this reading of *Rosehill*. There are five specific aspects of the book that suggest these ideas have relevance.

Mixing the social and the physical

Touring through the streets, Carol Lake mixes descriptions of the physical environment with those of people. It proved difficult to find a passage of physical description of the area with which to begin this section, as descriptions of people and place, local and distant events are so firmly bound up together. Walking down the street, the anonymous characters of the urban scene are animated by snippets of biography. The physical community becomes defined by the summation of individual biographies.

> On my way home from the pictures I pass Lesley and her man, sitting outside the supermarket in the dark, with a bottle. Have they left Mohammed's? They sit with their backs against the dim glass, perhaps staying close to what vestigial light there is in the empty store. I smile a useless smile through the night.
>
> For a while in the summer he walked about alone, looking lost without her, scurrying along with his head bent. Was she ill? Had she left? It was impossible to ask. So she's still around.
>
> In The Mafeking, Gail polishes glasses and smiles as one of the regulars tells her of the time he was stationed in Cyprus.

> Trisha, says Heather, has killed herself. Heather? The girl on the check-out in the chemist's. She was seventeen; she lived alone in a flat. She was very friendly at work but never went out at nights . . .
>
> The bulk of the pub on the corner, with its warm red brick and those windows whose upper panes are divided into little squares, it resembles a ship, and as I walk past it's floating, a big rosy galleon, up in the waves of the sky.
>
> Across the road Ron stands on his doorstep, tasting the air. He is waiting to hear from the hospital, to know whether or not he has a tumour. He sets out to collect a newspaper from the shop trying to enjoy each moment of not-knowing.
>
> (Lake 1989: 176)

Sharing a way of story-telling

For Carol Lake, individuals carry the accumulation of their identity; stories from the past are worn in our conversation like a suit of clothes. Talking of her neighbour, she says: 'As she talks, Hawa's different lives African, Liverpudlian, slide about her like iridescence on a bubble' (ibid.: 32). Even a documentary section relating a conversation with Ajit, an Indian neighbour, begins with the problems of her children at school in Derby and continues to tell of her student-day holidays in rural India, however, quite without change of tenor or tense, it contains material which relates to Indian folk legend. This turns a factual story into a folk story as might be told to children about a distant place or time, mixing one sense of reality with another derived from a distant culture whose traces run into and mix with those of the present. The fact that Carol Lake calls herself Carol Sing when writing in the *London Review of Books* suggests considerable sympathy with an Indian tradition of story-telling. Important here is that the very structure of narrative as a ritual and symbolic practice, a shared way of story-telling, expresses a communality between Carol Lake and the people of Rosehill.

If this is an efficacious means of generating a common link between herself and others in Rosehill, generating a sense of community, it is also something intensely personal. Creating herself from the fragments of the biographies of others, cleverly and consciously, Carol Lake casts doubt on the certainty of the past she wishes to cast aside. The power of creativity enables her to use apparent fantasy to shed doubt on what commonsense would tell us is most solid. The small boy Ahmed says:

> '. . . And in Pakistan there's devils, and they ride on your shoulder, and if you be naughty they pinch you, and – ew it's horrible.' He says with an accompanying wince of remembrance 'And there's hanging there. There is really. And this man in the next village, he was ever so very wicked, he made this other man dead all over 'cept for his eyeballs?'

'Dead all over . . . Don't tell lies.' I don't mind a few Polyfilla-lies, to staunch the gaps, as it were, and help the story along, but this seems a bit much.

I'm temporarily thrown by this, and just sit.

'Across the road there's this lady, and she goes mad in the summer, she goes up and down Rosehill Street breaking windows and taking her clothes off. And then some men come and put her in a van, and she goes off with them and they put wires on her head. That's what Shabir said.' I look at him. 'In King's Field,' he says 'why do they put wires on their heads?'

'I don't know. They just do. It makes them forget things; they forget their problems for a while. They forget who they are and everything that's ever happened to them.'

'*Now* who's telling lies?' says Ahmed, jubilant and pointing.

(Lake 1989: 78–9)

And it is hard for us to decide who, if anyone, is telling lies.

Community and personal estrangement

The parts of the book where a sense of community are most deeply explored are constructed around stories centred on explicitly spatial metaphors, which link people with place, past with present and future, invention with documentation.

Carol Lake is always aware of a similarity between her story and the story of others in Rosehill. Ahmed's or Ajit's memories of the Punjab and her own memories of a childhood Rosehill both represent the thoughts of aliens in an alienated land. Personal estrangement is a constant theme. Most important is the story of 'The Sisters', which comes near the middle of the book, bringing the stoic resistance of two elderly women hanging on to their home in the face of redevelopment plans together with a portrait of a devastated landscape. Carol remembers them from her childhood and decides to make some fudge and take Ahmed to see them in order to express sympathy and solidarity with their cause, at the same time bringing together the author's lives, past and present. Worried about what sort of reception she may receive, she stands at a distance and watches while Ahmed plays the role of ambassador and takes the fudge.

We pass through the arcade into a small street and across the main road, descending into an area of mud stretching down to the next main road. A small cluster of new flats clings to the road. Streets and streets have been wiped out, thousands of homes, houses with their own gardens, little trees, privacy. Mud, pools of dried concrete, the beginnings of a carpark – the whole area laid bare and the fall of the land exposed. The lower main road is windy and filthy with paper rubbish. St Andrew's House, a new Social Security office, dispensing Giros, not

> bread and wine, stands where the church once stood. We stand before it, maybe hoping for something to happen. Now the houses are down a fierce wind blasts through the road, claps papers, tosses my hair, makes Ahmed's trousers flap. He clutches his parka about him, and makes off with the fudge. I wait by the endless wall of the Social Security office. Over the road is the sisters' house, once just another house in a terrace, part of streets of terraced houses. Now it stands stark, the doorstep is well scrubbed, the dormer-window reflects the fading light; it is more remarkable than anything in sight. When Miss Hester stands at her doorway what does she see? Does she wipe away the ruins as she looks?
> (Lake 1989: 71)

He returns and she questions him about the sisters. Were they nice? Were they pleased to receive the fudge? What did they look like? Eventually, the little boy owns up that he never knocked on the door, he just dropped the gift and ran away. There is little communication and no reconciliation between past and present. Just as the sisters are imprisoned in a physical past isolated amid a sea of demolished houses, so Ahmed talks of a family in Pakistan his parents cannot afford to send him to visit, and Carol talks of a Rosehill past which is inaccessible to her. There is a deep and enduring estrangement which comes as a blow, a cleavage right at the heart of the book when it has disarmed us into thinking that it is at its most located. Situated deep within the very core of its rootedness, the stories people have as bonds of community inscribe only the boundaries of their personal loneliness.

Utopia and the worldliness of community

A story that examines the concept of community in Rosehill most explicitly is 'The Day of Judgement' which poses the question of what would happen if the people of Rosehill were faced with impending annihilation. If the story of the Sisters explores estrangement, then this explores reconciliation. The story uses the spatial metaphor of God visiting Rosehill in Noah's Ark on the day the world is engulfed by water. Surprisingly, God leaves the Ark and goes to the pub whilst the Ark leaves without him. As the Ark sails away:

> The inmates of the Ark gaze back over the side, some pityingly. Something funny has happened to them. They have gone flat, like pieces of paper.
>
> The Ark is on the edge of the horizon now, its destination the heartlessness of perfection. Most of the inmates already know what they are going to find – endless fruit, endless harmony, endless entropy, endless endless compassion, black and white in endless inane tableaux of equality. It sails off to a perfect world; the sky has turned into rich primary colours and in the distance the Ark bobs about on a bright blue sea.
>
> Below lie the streets, their corners and rough edges and pitted

> pavements prisming and splintering the rainbow's light into thousands and millions more lights, making the streets a thousand and million times more beautiful.
>
> (Lake 1989: 119)

Because Carol Lake portrays herself to a certain extent as a child of the 1960s, these passages have been interpreted by reviewers as the psychedelic excesses of a latter-day hippie. However, it is argued here that they refer back to the impossibility of escape explored in 'The Sisters'. Significantly, it focuses on a story about a journey, thereby referencing the stories of distant times and places which the people of Rosehill carry with them. Rather than celebrating an imagined Utopia, it suggests only the possibility of beauty in Rosehill where the fragmenting light playing off the rough worldliness of the broken streets transforms them, giving them a beauty a thousand times greater than the comic-book simplicity of the Ark sailing away to Utopia, with its cardboard cut-out figures sailing on a flat blue sea. The saved inhabit a world lacking depth, texture, dimensionality, contrast, roughness and smoothness, contradiction – the things that constitute a humanist humanity. She calls the travellers on the Ark inmates, and in doing so she expresses a fear of institutions which can be traced back to her days in the mental hospital where a clinical, clean, rule-governed existence threatens identity in a most profound sense. The sterility of Utopia is a very conservative notion and perhaps, like the impending possibility of oblivion, owes as much if not more to the 1980s than to the 1960s as one might believe from the imagery.

The Ark is a symbolic rallying point for the whole community, and she describes a whole series of different reactions to the coming of the Ark as a moment of collective sensibility. This reflects the claim that individuality and communality are constantly mediated within communities: at their most unified, the people of Rosehill express their greatest degree of individuality. Some go aboard the Ark and jeer at those who do not join; some stand by and jeer at those who go aboard; some carry on regardless as if to defy the whole idea of Armageddon; some turn away because salvation interferes with their day-to-day routine; some have their own saviour who apparently will come tomorrow; others prefer the pub.

But moral condemnation is reserved only for those who leave: remember God does not go with them, he stays in Rosehill, suggesting by his actions that the Kingdom of Heaven may be built on earth, that the chosen people may not necessarily be those who believe themselves to be chosen. Carol Lake's world is formed from the most powerless in an underprivileged society. One reviewer exclaims 'all her friends are either insane, foreign or under thirteen' (Clyde 1989: 804). In a very real way this reflects her status as a woman in a highly patriarchal local society which is itself deprived, where women are marginalized and constrained both because of their gender and because of their socio-economic status. So perhaps she is claiming salvation for those who have no voice, even in a society which has little to say in its own destiny. Lake is also expressing a deeply felt religious sensibility in saying that the beauty of a naturally created earth is better than that of an imagined cardboard heaven.

So Rosehill itself is a place of escape, a place of salvation where one can be oneself – have a room of one's own and write. 'Normal' morality is suspended or inverted and the expectations and social prescriptions of the outside world do not stifle one's sense of self.

Discussion: Community and identity

Lake used to be a political activist, yet it is the apparent amorality of these stories which confounds its reviewers:

> Secretly, we would rather she ranted, pathetically and with suitable political overtones about what she has seen. But this would still leave us comfortably separate from it all.
>
> (Clyde 1989: 804)

One only has to look at her views on Utopia and racial harmony, for example, to believe that the book signals disillusion and disenchantment, yet she creates a world which is full of expectation and enchantment. For Carol Lake, authoritarian attempts to generate community are fated by the stilted hand of bureaucracy, the children's play equipment in the Arboretum is an iron cage which imprisons creativity; the new corporate pub sign above the Falstaff turns him into a puritanical abstainer. This is not a claim for the enterprise of private capital, or even the egalitarian autonomy of local decision-making, but rather for an underdetermined world of loose ends where opportunity relies on the proliferation of margins. Carol Lake's Rosehill is not gendered simply. She moves in a 'man's world': local residents say 'single women do not live alone around here'. However distanced and disarmed in the character of the boy Ahmed, a dialogue with the patriarchal world is brought constantly into Carol Lake's most domestic moments. Amongst her friends is the publican Sassy, and together their worlds represent an uncompromising interweaving of public and private, female and male. She uses the anonymity engendered by the subordinate role of women in Rosehill positively in order to give herself a space without domestic responsibility in which to write. However, her stories are also about creating a home, about comfort and domesticity at the edge and under threat. Home is a shop doorway, a terrace in a sea of derelict ground, a street frequented by football thugs and drunks. Yet, after refusing to patronize the subjects of her stories by idealizing them, she concludes with a touch of nostalgic retrospection as she remembers the Rosehill of her childhood. For a moment she trys to justify Rosehill in the romantic manner understandable to outsiders, yet as if echoing the moral tones in 'The Day of Judgement', she does not seek to escape from its present and still lives in the district. The continued tussle with Western notions of reality in the stories, her reluctance to be pinned down as a person, her enjoyment of multiple identities, her refusal, in short, to be classified and categorized is a form of resistance. This is expressed geographically in the acts of mapping and naming. She says of Rosehill:

> This area is called by many names: Rosehill, Peartree, Normanton. My favourite is Litchurch, but this isn't used now. The areas all merge into one another here and there, and so it's good not to have a precise name, it's like the life here is still escaping. I'll hate it if any of the names become official, to the exclusion of others.
>
> (Lake 1989: 169)

If community is about creating certainty in an insecure world, then Carol Lake's Rosehill appears to fall down on this most fundamental of community qualities. The individuality of mass society has won out against the moral rectitude of traditional community. However, for Lake, certainty is not built in this way, stories build bridges between self and locality, certainty comes from possessing the means to describe oneself, and security comes from doing this in a way that is shared by the group and unavailable to outsiders. Her stories are spacious and her community is about creating a sense of space, rooted in the worldliness of locality and its everyday life. It is for this reason that the reviewer recognizes that we are uncomfortably close to Rosehill in spite of the absence of political rhetoric. Her celebration of ordinariness is important, where this may be described as a lack of direction, an ambiguity, something which is not trying to become something else, something which does not stand out from the crowd. This is not a sign of apathy, it is a deliberate tactical campaign, because apparent purposelessness renders Lake and Rosehill meaningless to the forces of authority. In this sense, Carol Lake's Rosehill is indeed a place which is secure and protected from the outside world, from the forces of classification, categorization and imposed progress which stifle creativity.

As one reviewer says: 'I can truly give it no higher praise than to say that if she read it and could *understand* it, Margaret Thatcher would hate it' (Parker 1989: 36–7).

NOTES

1 The terms 'self-identity' and 'personal identity' are used here in a rather loose way, without a thoroughly worked out theory of what the self might be. As it is used in this chapter self-identity relates to self-consciousness and the nexus of ideas that relate to being a person independent of other beings. It is therefore related to the processes of differentiation which separate self from other, it brings results to consciousness which are not necessarily wholly or even partly motivated by consciousness, and it has implications for autonomy and volitional action which are socially or ideologically determined.

The term 'story-telling' is used rather than 'narrative' in order to stress the partial, temporary and mediated quality of such biographical fragments.

2 As Wright puts it: 'The point should be clear enough. People live in different worlds even though they share the same locality. What is pleasantly "old" for one person is decayed and broken for another, just as a person with money has a different experience of shopping in the area than someone with almost none. A white home-owner is likely to have a different experience of the police than a black person – home-owner or not. Likewise, if I read the *Guardian* or *The Times* and can substantially determine my own relation to the borough, then maybe I don't actually need to read the *Hackney Gazette*. Those stories of daily misery and violent horror can stay local to someone else's paper, together with the job advertisements (although, of course, I'll keep a close eye on the rising house prices)' (Wright 1985: 237).

3 The concept of 'objectification' does not represent alienation, a Hegelian or Lukacs-esque reification or estrangement from the essence of humanity. It cannot do so where the self is viewed as socially constructed and where language is fundamental to our experience of 'reality'.

4 I do not necessarily agree with ideas implied by the concepts 'mass society' and 'post-industrial society' but use this because so much of the community studies literature in the 1960s and early 1970s was constructed in dialogue with these sociological currents. Newby and Bell rail against Marshall McLuhan: 'Another line of criticism of community studies that is increasingly persuasive is that in 'post-industrial society' (or some such cliché) all become so mobile that community has become irrelevant. Marshall McLuhan would have us all living in 'global villages' and now 'that the message can travel faster than the messenger' face-to-face communities are unnecessary' (Newby and Bell 1971: 18).

REFERENCES

Bakhtin, M. (1981) *The Dialogic Imagination: Four Essays*, ed. M. Holquist, Austin: University of Texas Press.

Barthes, R. (1977) *Image, Music, Text*, London: Fontana.

Baynes, K., Bohman, J. and McCarthy T. (1987) *After Philosophy: End or Transformation?*, Cambridge, Mass: MIT Press.

Bell, C. and Newby, H. (1971) *Community Studies: An Introduction to the Sociology of the Local Community*, London: Allen & Unwin.

Belsey, C. and Moore, J. (eds) (1989) *The Feminist Reader: Essays in Gender and the Politics of Literary Criticism*, London: Macmillan.

Bernasconi, R. and Critchley, S. (eds) (1991) *Re-reading Levinas*, Bloomington: Indiana University Press.

Bernasconi, R. and Wood, D. (eds) (1988) *The Provocation of Levinas: Rethinking the Other*, London: Routledge.

Braine, J. (1957) *Room at the Top*, London: Eyre & Spottiswoode.

Chambers, R. (1984) *Story and Situation: Narrative Seduction and the Power of Fiction*, Minneapolis: University of Minnesota Press.

Clapp, S. (1989) 'Coming out with something: review of *Rosehill Portraits from a Midlands City*', *London Review of Books* 11(13).

Clark, J. (1973) 'The concept of community: a re-examination', *Sociological Review* 21.

Clifford, J. (1988) *The Predicament of Culture: Twentieth Century Ethnography, Literature, and Art*, London: Harvard University Press.

Clyde, T. (1989) 'Review of Rosehill: portraits of a Midlands City', *Times Literary Supplement*, July 21–7: 804.

Cohen, A. P. (1982) *Belonging: Identity and Social Organisation in British Rural Communities*, Manchester: Manchester University Press.

——— (1985) *The Symbolic Construction of Community*, London: Tavistock.
——— (ed.) (1986) *Symbolising Boundaries: Identity and Diversity in British Cultures*, Manchester: Manchester University Press.
Colls, R. and Dodd, P. (eds) (1986) *Englishness: Politics and Culture 1880–1920*, London: Croom Helm.
de Certeau, M. (1984) *The Practice of Everyday Life*, Berkeley: University of California Press.
Deleuze, G. and Guattari, F. (1983) *Anti-Oedipus*, Minneapolis: University of Minnesota Press.
——— ——— (1988) *A Thousand Plateaus: Capitalism and Schizophrenia*, London: Athlone.
Dennis, N., Henriques, F. and Saughter, F. (1956) *Coal is Our Life*, London: Eyre & Spottiswoode.
Derrida, J. (1991) 'At this very moment in this work here I am', in Bernasconi and Wood (eds), 11–47.
Eyles, J. (1985) *Senses of Place*, Warrington.
Frankenberg, R. (1966) *Communities in Britain: Social Life in Town and Country*, Harmondsworth: Penguin.
Ganguly, K. (1992) 'Migrant identities: personal memory and the construction of selfhood', *Cultural Studies* 6(1): 27–50.
Gergan, K. J. and Davis, K. E. (eds) (1985) *The Social Construction of the Person*, New York: Springer-Verlag.
Giddens, A. (1991) *Modernity and Self-Identity: Self and Society in the Late Modern Age*, Oxford: Polity Press.
Godzich, W. (1986) 'The further possibility of knowledge' in M. de Certeau, *Heterologies: Discourse on the Other*, Manchester: Manchester University Press.
Harper, S. (1989) 'The British rural community: an overview of perspectives', *Journal of Rural Studies* 5(2): 161–84.
Harré, R. (1983) *Personal Being: A Theory for Individual Psychology*, Oxford: Blackwell.
Harvey, D. (1989a) *The Urban Experience*, Oxford: Blackwell.
——— (1989b) *The Condition of Postmodernity*, Oxford: Blackwell.
Hearn, J. (1987) *The Gender of Oppression: Men, Masculinity and the Critique of Marxism*, Brighton: Wheatsheaf.
Heller, A. (1984) *Everyday Life*, London: Routledge & Kegan Paul.
Hewison, R. (1987) *The Heritage Industry: Britain in a Climate of Decline*, London: Methuen.
Hoggart, R. (1957) *The Uses of Literacy*, Harmondsworth: Penguin.
hooks, b. (1991) *Yearning: Race, Gender and Cultural Politics*, London: Turnaround.
Jackson, P. (1991a) 'Mapping meanings: a cultural critique of locality studies', *Environment and Planning A* 23: 215–28.
——— (1991b) 'The cultural politics of masculinity: towards a social geography', *Transactions I.B.G.* N.S. 16: 199–213.
Jardine, A. and Smith, P. (1987) *Men in Feminism*, London: Methuen.
Kotre, J. (1984) *Outliving the Self*, Baltimore: Johns Hopkins University Press.
Laclau, E. and Mouffe, C. (1985) *Hegemony and Socialist Strategy: Toward a Radical Democratic Politics*, London: Verso.
Lake, C. (1989) *Rosehill: Portraits from a Midlands City*, London: Bloomsbury.
Lasch, C. (1985) *The Minimal Self: Psychic Survival in Troubled Times*, London: Picador.
Levinas, E. (1981) *Otherwise than Being or Beyond Essence*, The Hague: Martinus Nijhoff.
Lyotard, J.-F. (1988) *The Differend*, Minneapolis: University of Minnesota Press.
MacIntyre, A. (1981) *After Virtue: a Study in Moral Theory*, London: Duckworth.
Maclean, M. (1988) *Narrative as Performance: The Baudelairean Experiment*, London: Routledge.

Marcus, G. E. and Fischer, M. (1986) *Anthropology as Cultural Critique: An Experimental Moment in the Human Sciences*, London: University of Chicago Press.
Massey, D. (1991) 'The political place of locality studies', *Environment and Planning A* 23: 267–81.
Moi, T. (1985) *Sexual/Textual Politics: Feminist Literary Theory*, London: Methuen.
Newby, H. and Bell, C. (1971) *Community Studies, an Introduction to the Sociology of the Local Community*, London: Allen & Unwin.
Nisbet, R. (1962) *Community and Power*, London: Oxford University Press.
—— (1967) *The Sociological Tradition*, London: Heinemann.
Norris, C. (1987) *Jacques Derrida*, London: Fontana.
Parker, T. (1989) 'Friends and neighbours: review of *Rosehill: Portraits from a Midlands City*', *New Statesman and Society* 2, May 26: 36.
Plant, R., Lesser, H. and Taylor-Gooby, P. (1980) *Political Philosophy and Social Welfare: Essays on the Normative Basis of Welfare Provision*, London: Routledge & Kegan Paul.
Revill, G. (1991) 'Trained for life: personal identity and the meaning of work in the nineteenth-century railway industry', in C. Philo (ed.) *New Words, New Worlds: Reconceptualising Social and Cultural Geography: Conference Proceedings*, Lampeter: SCGS:
Ricoeur, P. (1971) 'The model of the text: meaningful action considered as a text', *Social Research* 38: 529–62.
—— (1981) *Hermeneutics and the Human Sciences*, Cambridge: Cambridge University Press.
Rimmon-Kenan, S. (1983) *Narrative Fiction: Contemporary Poetics*, London: Methuen.
Roberts, R. (1971) *The Classic Slum: Salford Life in the First Quarter of the Century*, Manchester: University of Manchester Press.
Rose, N. (1990) *Governing the Soul: The Shaping of the Private Self*, London: Routledge.
Rowe, D. (1982) *The Construction of Life and Death*, Chichester: John Wiley.
Said, E. (1978) *Orientalism*, London: Routledge & Kegan Paul.
Sennet, R. (1977) *The Fall of Public Man*, Cambridge: Cambridge University Press.
Sillitoe, A. (1958) *Saturday Night and Sunday Morning*, London: W. H. Allen.
—— (1988) *The Far Side of the Street: Fifteen Short Stories*, London: W. H. Allen.
Spivak, G. C. (1987) *In Other Worlds: Essays in Cultural Politics*, London: Methuen.
—— (1988) 'Can the subaltern speak?', in C. Nelson and L. Grossberg (eds) *Marxism and the Interpretation of Culture*, London: Macmillan.
Stacey, M. (1969) 'The myth of community studies', *British Journal of Sociology* 20: 134–47.
Taylor, C. (1989) *Sources of the Self: The Making of Modern Identity*, Cambridge: Cambridge University Press.
Tressell, R. (reprint 1955) *The Ragged Trousered Philanthropist*, London: Lawrence & Wishart.
Wallman, S. (1984) *Eight London Households*, London: Tavistock.
Wiener, M. J. (1981) *English Culture and the Decline of the Industrial Spirit, 1850–1980*, Cambridge: Cambridge University Press.
Williams, R. (1976) *Keywords*, London: Fontana.
Wood, M. R. and Zurchner, L. A. (1988) *The Development of a Postmodern Self*, New York: Greenwood.
Wright, P. (1985) *On Living in an Old Country: The National Past in Contemporary Britain*, London: Verso Books.
Young, R. (1990) *White Mythologies: Writing History and the West*, London: Routledge & Kegan Paul.

8

POLITICS AND SPACE/TIME[1]

Doreen Massey

'Space' is very much on the agenda these days. On the one hand, from a wide variety of sources come proclamations of the significance of the spatial in these times: 'It is space not time that hides consequences from us' (Berger); 'The difference that space makes' (Sayer); 'that new spatiality implicit in the post-modern' (Jameson); 'it is space rather than time which is the distinctively significant dimension of contemporary capitalism' (Urry); and 'All the social sciences must make room for an increasingly geographical conception of mankind' (Braudel). Even Foucault is now increasingly cited for his occasional reflections on the importance of the spatial. His 1967 Berlin lectures contain the unequivocal: 'The anxiety of our era has to do fundamentally with space, no doubt a great deal more than with time'. In other contexts the importance of the spatial, and of associated concepts, is more metaphorical. In debates around identity the terminology of space, location, positionality and place figures prominently. Homi Bhabha, in discussions of cultural identity, argues for a notion of a 'third space'. Jameson, faced with what he sees as the global confusions of postmodern times, 'the disorientation of saturated space', calls for an exercise in 'cognitive mapping'. And Laclau, in his own very different reflections on the 'new revolution of our time', uses the terms 'temporal' and 'spatial' as the major differentiators between ways of conceptualizing systems of social relations.

In some ways, all this can only be a delight to someone who has long worked as a 'geographer'. Suddenly the concerns, the concepts (or, at least, the *terms*) which have long been at the heart of our discussion are at the centre also of wider social and political debate.

And yet, in the midst of this gratification, I have found myself uneasy about the way in which, by some, these terms are used. Here I want to examine just one aspect of these anxieties about some of the current use of spatial terminology – the conceptualization (often implicit) of the term 'space' itself.

In part, this concern about what the term 'space' is meant to mean arises simply from the multiplicity of definitions adopted. Many authors rely heavily on the terms 'space'/'spatial', and each assumes that their meaning is clear and uncontested. Yet in fact the meaning which different authors assume (and therefore – in the case of metaphorical usage – the import of the metaphor) varies greatly. Buried in these unacknowledged disagreements is a debate which never surfaces; and it never surfaces because everyone assumes we already know what these terms mean. Henri Lefebvre, in the opening pages of his book *The Production of Space*, commented on just this phenomenon: the fact that authors who excel in logical rigour in so many ways will fail to define a term which functions crucially in their argument: 'Conspicuous by its absence from supposedly epistemological studies is . . . the idea . . . of space – the fact that 'space' is mentioned on every page notwithstanding' (Lefebvre 1991: 3). At least there ought to be a debate about the meaning of this much-used term.

None the less, had this been all, I would probably not have been exercised to write about it. But the problem runs more deeply than this. For among the many and conflicting definitions of space that are current in the literature, there are some – and very powerful ones – which deprive it of politics and of the possibility of politics: they effectively depoliticize the realm of the spatial. By no means all authors relegate space in this way. Many, drawing on terms such as 'centre', 'periphery', margin', etc. and examining the 'politics of location' for instance, think of spatiality in a highly active and politically enabling manner. But for others, space is the sphere of the lack of politics.

Precisely because the use of spatial terminology is so frequently unexamined, this use of the term is not always immediately evident. It dawned fully on me when I read a statement by Ernesto Laclau in his *New Reflections on the Revolution of our Time*: 'Politics and space', he writes, 'are antinomic terms. Politics only exist insofar as the spatial eludes us' (Laclau 1990: 68).[2] For someone who, as a geographer, has for years been arguing, along with many others, for a dynamic and politically progressive way of conceptualizing the spatial, this was clearly provocative!

Because my own enquiries were initially stimulated by Laclau's book, and because unearthing the implicit definitions at work implies a detailed reading (which restricts the number of authors who can be considered), this discussion takes *New Reflections* as a starting point, and considers it in most detail. But, as will become clear, the implicit definition used by Laclau, and which depoliticizes space, is shared by many other authors. In its simpler forms, it operates, for instance, in the debate over the nature of structuralism, and is an implicit reference point in many a text. It is moreover in certain of its fundamental aspects shared by authors, such as Fredric Jameson, who in other ways are making arguments very different from those of Laclau.

To caricature it rather crudely, Laclau's view of space is that it is the realm of stasis. There is, in the realm of the spatial, no true temporality and thus no possibility of politics. It is on this view, and on a critique of it, that much of the initial discussion will be concentrated. But in other parts of the debate about the nature of the current era, and in particular in relation to 'postmodernity', the realm of the spatial is given entirely different associations from those ascribed by Laclau. Thus Jameson, who sees postmodern times as being particularly characterized by the importance of spatiality, interprets it in terms of an unnerving multiplicity: space is chaotic depthlessness (Jameson 1991). This is the opposite of Laclau's characterization, yet for Jameson it is – once again – a formulation that deprives the spatial of any meaningful politics.

A caveat must be entered from the start. This discussion will be addressing only one aspect of the complex realm which goes by the name of the spatial. Lefebvre, among others, insisted on the importance of considering not only what might be called 'the geometry' of space but also its lived practices and the symbolic meaning and significance of particular spaces and spatializations. Without disagreeing with that, the concentration here will none the less be on the view of space as what I shall provisionally call 'a dimension'. The argument is that different ways of conceptualizing this aspect of 'the spatial' themselves provide very different bases (or in some cases no basis at all) for the politicization of space. Clearly, anyway, the issue of the conceptualization of space is of more than technical interest; it is one of the axes along which we experience and conceptualize the world.

SPACE AND TIME

An examination of the literature reveals, as might be expected, a variety of uses and meanings of the term 'space', but there is one characteristic of these meanings which is particularly strong and widespread. This is the view of space which, in one way or another, defines it as stasis, and as utterly opposed to time. Laclau, for whom the contrast between what he labels temporal and what he calls spatial is key to his whole argument, uses a highly complex version of this definition. For him, notions of time and space are related to contrasting methods of understanding social systems. In his *New Reflections on the Revolution of our Time*, Laclau posits that 'any repetition that is governed by a structural law of successions is space' (Laclau 1990: 41) and 'spatiality means coexistence within a structure that establishes the positive nature of all its terms' (ibid.: 69). Here, then, any postulated causal structure which is complete and self-determining is labelled 'spatial'. This does not mean that such a 'spatial' structure cannot change – it may do – but the essential characteristic is that all the causes of any change which may take place are internal to the structure itself. On this view, in the realm of the spatial there can be no surprises (provided that we are analytically well equipped). In contrast to the closed and self-determining systems of the spatial, time (or temporality) for Laclau takes the form of dislocation, a dynamic which disrupts the predefined terms of any system of causality. The spatial, because it lacks dislocation, is devoid of the possibility of politics.

This is an importantly different distinction between time and space from that which simply contrasts change with an utter lack of movement. In Laclau's version, there can be movement and change within a so-called spatial system; what there cannot be is real dynamism in the sense of a change in the terms of 'the system' itself (which can therefore never be a simply coherent closed system). A distinction is postulated, in other words, between different types of what would normally be called time. On the one hand, there is the time internal to a closed system, where things may change yet without really changing. On the other hand, there is genuine dynamism, Grand Historical Time. In the former is included cyclical time, the times of reproduction, the way in which a peasantry represents to itself (says Laclau ibid.: 42) the unfolding of the cycle of the seasons, the turning of the earth. To some extent, too, there is 'embedded time', the time in which our daily lives are set.[3] These times, says Laclau, this kind of 'time' is space.

Laclau's argument here is that what we are inevitably faced with in the world are 'temporal' (by which he means dislocated) structures: dislocation is intrinsic and it is this – this essential openness – which creates the possibility of politics. Any attempt to represent the world 'spatially', including even the world of physical space, is an attempt to ignore that dislocation. Space therefore, in his terminology, is representation, is any (ideological) attempt at closure: 'Society, then, is unrepresentable: any representation – *and thus any space* – is an attempt to constitute society, not to state what it is' (Laclau 1990: 82, my emphasis). Pure spatiality, in these terms, cannot exist: 'The ultimate failure of all hegemonisation [in Laclau's term, spatialization], then, means that the real – including physical space – is in the ultimate instance temporal' (ibid.: 42); or again: 'the mythical nature of any space' (ibid.: 68). This does not mean that the spatial is unimportant. This is not the point at issue, nor is it Laclau's intent. For the 'spatial' as the ideological/mythical is seen by him as itself part of the social and as constitutive of it: 'And insofar as the social is impossible without some fixation of meaning, without the discourse of closure, the ideological must be seen as constitutive of the social' (ibid.: 92).[4] The issue here is not the relative priority of the temporal and the spatial, but their definition. For it is through this logic, and its association of ideas with temporality and spatiality, that Laclau arrives at the depoliticization of space. 'Let us begin', writes Laclau, 'by identifying three dimensions of the relationship of dislocation that are crucial to our analysis. The *first* is that dislocation is the very form of temporality. And temporality must be conceived as the exact opposite of space. The "spatialization" of an event consists of eliminating its temporality' (ibid.: 41, my emphasis).

The second and third dimensions of the relationship of dislocation (see above) take the logic further: 'The *second* dimension is that dislocation [which, remember, is the antithesis of the spatial] is the very form of possibility' and 'The *third* dimension is that dislocation is the very form of freedom. Freedom is the absence of determination' (ibid.: 42, 43; my emphases). This leaves the realm of the spatial looking like unpromising territory for politics. It is lacking in dislocation, the very form of possibility (the form of temporality), which is also 'the very form of freedom'. Within the spatial, there is only determination, and hence no possibility of freedom or of politics.

Laclau's characterization of the spatial is, however, a relatively sophisticated version of a much more general conception of space and time (or spatiality and temporality). It is a conceptualization in which the two are opposed to each other, and in which time is the one which matters and of which History (capital H) is made. Time Marches On but space is a kind of stasis, where nothing really happens.

There are a number of ways in which, it seems to me, this manner of characterizing space and the realm of the spatial is questionable. Three of them, chosen precisely because of their contrasts, because of the distinct light they each throw on the problems of this view of space, will be examined here. The first draws on the debates which have taken place in 'radical geography' over the last two decades and more; the second examines the issue from the point of view of a concern with gender; and the third examines the view from physics.

Radical geography

In the 1970s the discipline of geography experienced the kinds of developments described by Anderson in 'A culture in contraflow' (Anderson 1991) for other social sciences. The previously hegemonic positivist 'spatial science' was increasingly challenged by a new generation of Marxist geographers. The argument turned intellectually on how 'the relation between space and society' should be conceptualized. To caricature the debate, the spatial scientists had posited an autonomous sphere of the spatial in which 'spatial relations' and 'spatial processes' produced spatial distributions. The geography of industry, for instance, would be interpreted as simply the result of 'geographical location factors'. Countering this, the Marxist critique was that all these so-called spatial relations and spatial processes were actually social relations taking a particular geographical form. The geography of industry, we argued, could therefore not be explained without a prior understanding of the economy and of wider social and political processes. The aphorism of the 1970s was 'space is a social construct'. That is to say – though the point was perhaps not made clearly enough at the time – space is constituted through social relations and material social practices.

But this, too, was soon to seem an inadequate characterization of the social/spatial relation. For while it is surely correct to argue that space is socially constructed, the one-sidedness of that formulation implied that geographical forms and distributions were simply outcomes, the end-point of social explanation. Geographers would thus be the cartographers of the social sciences, mapping the outcomes of processes which could only be explained in other disciplines – sociology, economics, and so forth. What geographers mapped – the spatial form of the social – was interesting enough, but it was simply an end-product: it had no material effect. Quite apart from any demeaning disciplinary implications, this was plainly not the case. The events taking place all around us in the 1980s – the massive spatial restructuring both intranationally and internationally as an integral part of the social and economic changes – made it plain that, in one way or another, 'geography matters'. And so, to the aphorism of the 1970s – that space is socially constructed – was added in the 1980s the other side of the coin: that the social is spatially constructed too, and that makes a difference. In other words, and in its broadest formulation, society is necessarily constructed spatially, and that fact – the spatial organization of society – makes a difference to how it works.

But if spatial organization makes a difference to how society works and how it changes, then, far from being the realm of stasis, space and the spatial are also implicated (*contra* Laclau) in the production of history – and thus, potentially, in politics. This was not an entirely new thought; Henri Lefebvre, writing in 1974, was beginning to argue a very similar position:

> The space of capitalist accumulation thus gradually came to life, and began to be fitted out. This process of animation is admiringly referred to as history, and its motor sought in all kinds of factors: dynastic interests, ideologies, the ambitions of the mighty, the formation of nation states, demographic pressures, and so on. This is the road to a ceaseless analysing of, and searching for, dates and chains of events. Inasmuch as space is the locus of all such chronologies, might it not constitute a principle of explanation at least as acceptable as any other?
> (Lefebvre 1991: 275)

This broad position – that the social and the spatial are inseparable and that the spatial form of the social has causal effecticity – is now accepted increasingly widely, especially in geography and sociology,[5] though there are still those who would disagree, and beyond certain groups even the fact of a debate over the issue seems to have remained unrecognized (Anderson, for example, does not pick it up in his survey).[6] For those familiar with the debate, and who saw in it an essential step towards the politicization of the spatial, formulations of space as a static resultant without any effect – whether the simplistic versions or the more complex definitions such as Laclau's – seem to be very much a retrograde step.

However, in retrospect, even the debates within radical geography have still fully to take on board the implications of our own arguments for the way in which space might be conceptualized.

Issues of gender

For there are also other reservations, from completely different directions, which can be levelled against this view of space and which go beyond the debate which has so far taken place within radical geography. Some of these reservations revolve around issues of gender.

First of all, this manner of conceptualizing space and time takes the form of a dichotomous dualism. It is neither a simple statement of difference (A, B, . . .) nor a dualism constructed through an analysis of the interrelations between the objects being defined (capital : labour). It is a dichotomy specified in terms of a presence and an absence; a dualism which takes the classic form of A/not-A. As was noted earlier, one of Laclau's formulations of a definition is: 'temporality must be conceived as the exact opposite of space' (Laclau 1990: 41). Now, apart from any reservations which may be raised in the particular case of space and time (and which we shall come to later), the mode of thinking which relies on irreconcilable dichotomies of this sort has, in general, recently come in for widespread criticism. All the strings of these kinds of opposition with which we are so accustomed to work (mind–body; nature–culture; reason–emotion; and so forth) have been argued to be at heart problematical and a hindrance to either understanding or changing the world. Much of this critique has come from feminists.[7]

The argument is twofold. First, and less importantly here, it is argued that this way of approaching conceptualization is – in Western societies and more generally in societies where child-rearing is performed overwhelmingly by members of one sex (women) – more typical of males than of females. This is an argument which generally draws on object-relations theory approaches to identity-formation. Second, however, and of more immediate significance for the argument being constructed here, it has been contended that this kind of dichotomous thinking, together with a whole range of the sets of dualisms which take this form (we shall look at some of these in more detail below), are related to the construction of the radical distinction between genders in our society, to the characteristics assigned to each of them, and to the power relations maintained between them. Thus, Nancy Jay, in an article entitled 'Gender and dichotomy', examines the social conditions and consequences of the use of logical dichotomy (Jay 1981). She argues not only that logical dichotomy and radical gender distinctions are associated but also, more widely, that such a mode of constructing difference works to the advantage of certain (dominant) social groups:

> that almost any ideology based on A/Not-A dichotomy is effective in resisting change. Those whose understanding of society is ruled by such ideology find it very hard to conceive of the possibility of alternative forms of social order (third possibilities). Within such thinking, the only alternative to the *one* order is disorder.
>
> (Jay 1981: 54)

Genevieve Lloyd, too, in a sweeping history of 'male' and 'female' in Western philosophy, entitled *The Man of Reason* (Lloyd 1984), argues that such dichotomous conceptualizations, and – what we shall come to later – the prioritization of one term in the dualism over the other, is not only central to much of the formulation of concepts with which Western philosophy has worked but that it is dependent upon, and is instrumental in the conceptualization of, among other things, a particular form of radical distinction between female and male genders. Jay argues that 'Hidden, taken for granted, A/Not-A distinctions are dangerous, and because of their peculiar affinity with gender distinctions, it seems important for feminist theory to be systematic in recognizing them' (Jay 1981: 47). The argument is that the definition of 'space' and 'time' under scrutiny here is precisely of this form, and on that basis alone warrants further critical investigation.

But there is a further point. For within this kind of conceptualization, only one of the terms (A) is defined positively. The other term (not-A) is conceived only in relation to A, and as lacking in A. A fairly thorough reading of some of the recent literature which uses the terminology of space and time, and which employs this form of conceptualization, leaves no doubt that it is time which is conceived of as in the position of 'A', and space which is 'not-A'. Over and over again, time is defined by such things as change, movement, history, dynamism; while space, rather lamely by comparison, is simply the absence of these things. There are two aspects of this. First, this kind of definition means that it is time, and the characteristics associated with time, which are the primary constituents of both space and time; time is the nodal point, the privileged signifier. And, second, this kind of definition means that space is defined by absence, by lack. This is clear in the simple (and often implicit) definitions (time equals change/movement, space equals the lack of these things), but it can also be argued to be the case with more complex definitions such as those put forward by Laclau. For although in a formal sense it is the spatial which in Laclau's formulation is complete and the temporal which marks the lack (the absence of representation, the impossibility of closure), in the whole tone of the argument it is in fact space which is associated with negativity and absence. Thus: 'temporality must be conceived as the exact opposite of space. The "spatialization" of an event consists of eliminating its temporality' (Laclau 1990: 41).

Now, of course, in current Western culture, or in certain of its dominant theories, woman too is defined in terms of lack. Nor, as we shall see, is it entirely a matter of coincidence that space and the feminine are frequently defined in terms of dichotomies in which each of them is most commonly defined as not-A. There is a whole set of dualisms whose terms are commonly aligned with time and space. With Time are aligned History, Progress, Civilization, Science, Politics and Reason, portentous things with gravitas and capital letters. With space, on the other hand, are aligned the other poles of these concepts: stasis, ('simple') reproduction, nostalgia, emotion, aesthetics, the body. All these dualisms, in the way that they are used, suffer from the criticisms made above of dichotomies of this form: the problem of mutual exclusivity and of the consequent impoverishment of both of their terms. Other dualisms could be added which also map on to that between time and space. Jameson, for instance, as does a whole line of authors before him, clearly relates the pairing to that between transcendence and immanence, with the former connotationally associated with the temporal and immanence with the spatial. Indeed, in this and in spite of their other differences, Jameson and Laclau are very similar. Laclau's distinction between the closed, cyclical time of simple reproduction (spatial) and dislocated, changing history (temporal), even if the latter has no inevitability in its progressive movement, is precisely that. Jameson, who bemoans what he characterizes as the tendency towards immanence and the flight from transcendence of the contemporary period, writes of 'a world peculiarly without transcendence and without perspective . . . and indeed without plot in any traditional sense, since all choices would be equidistant and on the same level' (Jameson 1991: 269), and this is a world where, he believes, a sense of the temporal is being lost and the realm of the spatial is taking over.

Now, as has been pointed out many times, these dualisms which so easily map on to each other also map on to the constructed dichotomy between female and male. From Rousseau's seeing woman as a potential source of disorder, as needing to be tamed by Reason, to Freud's famous pronouncement that woman is the enemy of civilization, to the many subsequent critics and analysts of such statements of the 'obviousness' of dualisms, of their interrelation one with another, and of their connotations of male and female, such literature is now considerable.[8] And space, in this system of interconnected dualisms, is coded female. ' "Transcendence", in its origins, is a transcendence *of* the feminine', writes Lloyd (1984: 101), for instance. Moreover, even where the transcodings between dualisms have an element of inconsistency, this rule still applies. Thus where time is dynamism, dislocation and History, and space is stasis, space is coded female and denigrated. But where space is chaos (which you would think was quite different from stasis; more indeed like dislocation), then time is Order . . . and space is *still* coded female, only in this context interpreted as threatening.

Elizabeth Wilson, in her book *The Sphinx in the City* (Wilson 1991), analyses this latter set of connotations. The whole notion of city culture, she argues, has been developed as one pertaining to men. Yet within this context women present a threat, and in two ways. First, there is the fact that in the metropolis we are freer, in spite of all the also-attendant dangers, to escape the rigidity of patriarchal social controls which can be so powerful in a smaller community. Second and following from this, 'women have fared especially badly in western visions of the metropolis because they have seemed to represent disorder. There is fear of the city as a realm of uncontrolled and chaotic sexual licence, and the rigid control of women in cities has been felt necessary to avert this danger' (Wilson 1991: 157). 'Woman represented feeling, sexuality and even chaos, man was rationality and control' (ibid.: 87). Among male modernist writers of the early twentieth century, she argues – and with the exception of Joyce – the dominant response to the burgeoning city was to see it as threatening, while modernist women writers (Woolf, Richardson) were more likely to exult in its energy and vitality. The male response was perhaps more ambiguous than this, but it was certainly a mixture of fascination and fear. There is an interesting parallel to be drawn here with the sense of panic in the midst of exhilaration which seems to have overtaken some writers at what they see as the ungraspable (and therefore unbearable) complexity of the postmodern age. And it is an ungraspability seen persistently in spatial terms, whether through the argument that it is the new (seen-to-be-new) time–space compression, the new global–localism, the breaking-down of borders, which is the cause of it all, or through the interpretation of the current period as somehow in its very character intrinsically more spatial than previous eras. In Jameson, these two positions are brought together, and he displays the same ambivalence. He writes of 'the horror of multiplicity' (Jameson 1991: 363), of 'all the web threads flung out beyond my "situation" into the unimaginable synchronicity of other people' (ibid.: 362). It is hard to resist the idea that Jameson's (and others') apparently vertiginous terror (a phrase they often use themselves) in the face of the complexity of today's world (conceived of as social but also importantly as spatial) has a lot in common with the nervousness of the male modernist, nearly a century ago, when faced with the big city.

It is important to be clear about what is being said of this relationship between space/time and gender. It is not being argued that this way of characterizing space is somehow essentially male; there is no essentialism of feminine/masculine here. Rather, the argument is that the dichotomous characterization of space and time, along with a whole range of other dualisms which have been briefly referred to, and with their connotative interrelations, may both reflect and be part of the constitution of, among other things, the masculinity and femininity of the sexist society in which we live. Nor is it being argued that space should simply be reprioritized to an equal status with, or instead of, time. The latter point is important because there have been a number of contributions to the debate recently which have argued that, especially in modernist (including Marxist) accounts, it is time which has been considered the more important. Ed Soja, particularly in his book *Postmodern Geographies* (Soja 1989), has made an extended and persuasive case to this effect (although see the critique by Gregory (1990)). The story told earlier of Marxism within geography – supposedly the spatial discipline – is indicative of the same tendency. In a completely different context, Terry Eagleton has written in his introduction to Kristin Ross's *The Emergence of Social Space* (Ross 1988) that 'Ross is surely right to claim that this idea [the concept of space] has proved of far less glamorous appeal to radical theorists than the apparently more dynamic, exhilarating notions of narrative and history' (ibid.: xii). It is interesting to speculate on the degree to which this deprioritization might itself have been part and parcel of the system of gender connotations. Ross herself writes, 'The difficulty is also one of vocabulary, for while words like "historical" and "political" convey a dynamic of intentionality, vitality, and human motivation, "spatial", on the other hand, connotes stasis, neutrality, and passivity' (ibid.: 8), and in her analysis of Rimbaud's poetry and of the nature of its relation to the Paris Commune she does her best to counter that essentially negative view of spatiality. (Jameson, of course, is arguing pretty much the same point about the past prioritization of time, but his mission is precisely the opposite of Ross's and Soja's: it is to hang on to that prioritization.)

The point here, however, is not to argue for an upgrading of the status of space within the terms of the old dualism (a project which is arguably inherently difficult anyway, given the terms of that dualism), but to argue that what must be overcome is the very formulation of space/time in terms of this kind of dichotomy. The same point has frequently been made by feminists in relation to other dualisms, most particularly perhaps – because of the debate over the writings of Simone de Beauvoir – the dualism of transcendence and immanence. When de Beauvoir wrote:

> Man's design is not to repeat himself in time: it is to take control of the instant and mould the future. It is male activity that in creating values has made of existence itself a value; this activity has prevailed over the confused forces of life; it has subdued Nature and Woman
>
> (de Beauvoir 1972: 97)

she was making precisely that distinction between cyclicity and 'real change' which not only is central to the classic distinction between immanence and transcendence but is also part of the way in which Laclau distinguishes between what he calls the spatial and the temporal. De Beauvoir's argument was that women should grasp the transcendent. A later generation of feminists has argued that the problem is the nature of the distinction itself. The position here is both that the two dualisms (immanence/transcendence and space/time) are related and that the argument about the former dualism could and should be extended to the latter. The next line of critique, the view from physics, provides some further hints about the directions which that reformulation might take.

The view from physics

The conceptualization of space and time under examination here also runs counter to notions of space and time within the natural sciences, and most particularly in physics. Now, in principle, this may not be at all important; it is not clear that strict parallels can or should be drawn between the physical and the social sciences. And indeed there continue to be debates on this subject in the physical sciences. The point is, however, that the view of space and time outlined above already does have, as one of its roots at least, an interpretation drawn – if only implicitly – from the physical sciences. The problem is that it is an outmoded one.

The viewpoint, as used for instance by Laclau, accords with the viewpoint of classical, Newtonian physics. In classical physics, both space and time exist in their own right, as do objects. Space is a passive arena, the setting for objects and their interaction. Objects, in turn, exist prior to their interactions and affect each other through force-fields. The observer, similarly, is detached from the observed world. In modern physics, on the other hand, the identity of things is *constituted through* interactions. In modern physics, while velocity, acceleration and so forth are defined, the basic ontological categories, such as space and time, are not. Even more significantly from the point of view of the argument here, in modern physics, physical reality is conceived of as a 'four-dimensional existence instead of . . . the evolution of a three-dimensional existence' (Stannard 1989). Thus, 'According to Einstein's theory . . . space and time are not to be thought of as separate entities existing in their own right – a three-dimensional space, and a one-dimensional time. Rather, the underlying reality consists of a four-dimensional space-time' (ibid.: 35). Moreover the observer, too, is part of the observed world.

It is worth pausing for a moment to clarify a couple of points here. The first point is that the argument here is not in favour of a total collapse of the differences between something called the spatial and the temporal dimensions. Nor, indeed, would that seem to be what modern physics is arguing either. Rather, the point is that space and time are inextricably interwoven. It is not that we cannot make any distinction at all between them, but that the distinction we *do* make needs to hold the two in tension, and to do so within an overall, and strong, concept of four-dimensionality.

The second point is that the definitions of both space and time in themselves must be constructed as the result of interrelations. This means that there is no question of defining space simply as not-time. It must have a positive definition, in its own terms, just as does time. Space must not be consigned to the position of being conceptualized in terms of absence or lack. It also means, if the positive definitions of both space and time must be interrelational, that there is no absolute dimension: space. The existence of the spatial depends on the interrelations of objects: 'In order for "space" to make an appearance there needs to be at least two fundamental particles' (ibid.: 33). This is, in fact, saying no more than what is commonly argued, even in the social sciences – that space is not absolute, it is relational. Perhaps the problem at this point is that the implications of this position seem not to have been taken on board.

Now, in some ways all this does seem to have some similarities with Laclau's use of the notion of the spatial, for his definition does refer to forms of social interaction. As we have seen, however, he designates them (or the concepts of them) as spatial only when they form a closed system, where there is a lack of dislocation which can produce a way out of the postulated (but impossible) closure. However, such use of the term is anyway surely metaphorical. What it represents is evidence of the connotations which are being attached to the terms space and spatial. It is not directly talking of 'the spatial' itself. Thus, to take up Laclau's usage in more detail: at a number of points, as we have seen, he presents definitions of space in terms of possible (in fact, he would argue, impossible) causal structures: 'any repetition that is governed by a structural law of successions is space' (Laclau 1990: 41); or 'spatiality means coexistence within a structure that establishes the positive nature of all its terms' (Laclau 1990: 69). My question of these definitions and of other related ones, both elsewhere in this book and more widely – for instance in the debate over the supposed 'spatiality' of structuralism – is 'says who?' Is not this appellation in fact pure assertion? Laclau agrees in rejecting the possibility of the actual existence of pure spatiality in the sense of undislocated stasis. A further question must therefore be: why postulate it? Or, more precisely, why postulate it as 'space'? As we have just seen, an answer that proposes an absolute spatial dimension will not do. An alternative answer might be that this ideal pure spatiality, which only exists as discourse/myth/ideology is in fact a (misjudged) metaphor. In this case it is indeed defined by interrelations – this is certainly not 'absolute space', the independently existing dimension – and the interrelations are those of a closed system of social relations, a system outside of which there is nothing and in which nothing will dislocate (temporalize) its internally regulated functioning. But then my question is: why call it space? The use of the term 'spatial' here would seem to be purely metaphorical. In so far as such systems do exist – and even in so far as they are merely postulated as an ideal – they can in no sense *be* simply spatial nor exist only *in* space. In themselves they *constitute* a particular form of space-time.[9]

Moreover, as metaphors, the sense of Laclau's formulations goes against what I understand by – and shall argue below would be more helpful to understand by – space/the spatial. 'Any repetition that is governed by a structural law of successions'? – but *is* space so governed? As was argued above, radical geographers reacted strongly in the 1970s precisely against a view of 'a spatial realm', a realm, posited implicitly or explicitly by a wide range of then-dominant practitioners, from mathematicized 'regional scientists' to data-bashers armed with ferociously high regression coefficients, in which there were spatial processes, spatial laws and purely spatial explanations. In terms of causality, what was being argued by those of us who attacked this view was that the spatial is externally determined. A formulation like the one above, because of the connotations it attaches to the words space/spatial in terms of the nature of causality, thus takes us back a good two decades. Or again, what of the second of Laclau's definitions given above – that the spatial is the 'coexistence within a structure that establishes the positive nature of all its terms'? What then of the paradox of simultaneity and the causal chaos of happenstance juxtaposition which are, as we shall argue below (and as Jameson sees), integral characteristics of relational space?

In this procedure, any sort of stasis (for instance a self-regulating structural coherence which cannot lead to any transformation outside of its own terms) gets called space/spatial. But there is no reason for this save the prior definition of space as lacking in (this kind of) transformative dynamic *and*, equally importantly, an assumption that anything lacking in (this kind of) dynamism is spatial. Instead, therefore, of using the terms space (and time) in this metaphorical way to refer to such structures, why do we not remain with definitions (such as dislocated/undislocated) which refer to the nature of the causal structures themselves? Apart from its greater clarity, this would have the considerable advantage of leaving us free to retain (or maybe it is to develop) a more positive concept of space.

Indeed, conceptualizing space and time more in the manner of modern physics would seem to be consistent with Laclau's general argument. His whole point about radical historicity is this: 'any effort to spatialize time ultimately fails and space itself becomes an event' (Laclau 1990: 84). Spatiality in this sense is agreed to be impossible. ' "Articulation" . . . is the primary ontological level of the constitution of the real', writes Laclau (ibid.: 184). This is a fundamentally important statement, and one with which I agree. The argument here is thus not opposed to Laclau; rather it is that exactly the same reasoning, and manner of conceptualization, which he applies to the rest of the world should be applied to space and time as well. It is not that the interrelations between objects occur *in* space and time; it is these relationships themselves which *create/define* space and time (Stannard 1989: 33).

It is not, of course, necessary for the social sciences simply to follow the natural sciences in such matters of conceptualization.[10] In fact, however, the views of space and time which are being examined here do, if only implicitly, tend to lean on versions of the world derived from the physical sciences; but the view they rely on is one which has been superseded theoretically. Even so, it is still the case that even in the natural sciences it is possible to use different concepts/ theories for different purposes. Newtonian physics is still perfectly adequate for building a bridge. Moreover there continue to be debates between different parts of physics. What is being argued here is that the social issues that we currently need to understand, whether they be the high-tech postmodern world or questions of cultural identity, require something that would look more like the 'modern physics' view of space. It would, moreover, precisely by introducing into the concept of space that element of dislocation/freedom/ possibility, enable the politicization of space/space-time.

An alternative view of space

A first requirement of developing an alternative view of space is that we should try to get away from a notion of society as a kind of 3-D (and indeed more usually 2-D) slice which moves through time. Such a view is often, even usually, implicit rather than explicit, but it is remarkably pervasive. It shows up in the way people phrase things, in the analogies they use. Thus, just briefly to cite two of the authors who have been referred to earlier, Foucault writes: 'We are at a moment, I believe, when our experience of the world is less that of a long life developing through time than that of a network that connects points and intersects with its own skein' (Foucault 1986: 22), and Jameson contrasts 'historiographic deep space or perspectival temporality' with a (spatial) set of connections which 'lights up like a nodal circuit in a slot machine' (Jameson 1991: 374). The aim here is not to disagree in total with these formulations, but to indicate what they imply. What they both point to is a contrast between temporal movement on the one hand and, on the other hand, a notion of space as instantaneous connections between things at one moment. For Jameson, the latter type of (inadequate) history-telling has replaced the former. And if this is true it is indeed inadequate. But while the contrast – the shift in balance – to which both authors are drawing attention is a valid one, in the end the notion of space as *only* systems of simultaneous relations, the flashing of a pin-ball machine, is inadequate. For, of course, the temporal movement is also spatial; the moving elements have spatial relations to each other. And the 'spatial' interconnections which flash across can only be constituted temporally as well. Instead of linear process counterposed to flat surface (which anyway reduces space from three to two dimensions), it is necessary to insist on the irrefutable four-dimensionality (indeed 'n'-dimensionality) of things. Space is not static, nor time spaceless. Of course, spatiality and temporality are different from each other but neither can be conceptualized as the absence of the other. The full implications of this will be elaborated below, but for the moment the point is to try to think in terms of all the dimensions of space-time. It is a lot more difficult than at first sight it might seem.

Second, we need to conceptualize space as constructed out of interrelations, as the simultaneous coexistence of social interrelations and interactions at all spatial scales, from the most local level to the most global. Earlier it was reported how, in human geography, the recognition that the spatial is socially constituted was followed by the perhaps even more powerful (in the sense of the breadth of its implications) recognition that the social is necessarily spatially constituted too. Both points (though perhaps in reverse order) need to be grasped at this moment. On the one hand, all social (and indeed physical) phenomena/activities/relations have a spatial form and a relative spatial location: the relations that bind communities, whether they be 'local' societies or worldwide organizations; the relations within an industrial corporation; the debt relations between the South and the North; the relations that result in the current popularity in European cities of music from Mali. The spatial spread of social relations can be intimately local or expansively global, or anything in between. Their spatial extent and form also change over time (and there is considerable debate about what is happening to the spatial form of social relations at the moment). But, whatever way it is, there is no getting away from the fact that the social is inexorably also spatial.

The proposition here is that this fact be used to define the spatial. Thus, the spatial is socially constituted. 'Space' is created out of the vast intricacies, the incredible complexities, of the interlocking and the non-interlocking, and the networks of relations at every scale from local to global. What makes a particular view of these social relations specifically spatial is their simultaneity. It is a simultaneity, also, which has extension and configuration. But simultaneity is absolutely not stasis. Seeing space as a moment in the intersection of configured social relations (rather than as an absolute dimension) means that it cannot be seen as static. There is no choice between flow (time) and a flat surface of instantaneous relations (space). Space is not a 'flat' surface in that sense because the social relations which create it are themselves dynamic by their very nature. It is a question of a manner of thinking. It is not the 'slice through time' which should be the dominant thought but the simultaneous coexistence of social relations that cannot be conceptualized as other than dynamic. Moreover, and again as a result of the fact that it is conceptualized as created out of social relations, space is by its very nature full of power and symbolism, a complex web of relations of domination and subordination, of solidarity and co-operation. This aspect of space has been referred to elsewhere as a kind of 'power-geometry' (Massey 1993).

Third, this in turn means that the spatial has *both* an element of order *and* an element of chaos (or maybe it is that we should question that dichotomy also). It cannot be defined on one side or the other of the mutually exclusive dichotomies discussed earlier. Space has order in two senses. First, it has order because all spatial locations of phenomena are caused; they can in principle be explained. Second, it has order because there are indeed spatial systems, in the sense of sets of social phenomena in which spatial arrangement (that is, mutual relative positioning rather than 'absolute' location) itself is part of the constitution of the system. The spatial organization of a communications network, or of a supermarket chain with its warehousing and distribution points and retail outlets, would both be examples of this, as would the activity space of a multinational company. There is an integral spatial coherence here, which constitutes the geographical distributions and the geographical form of the social relations. The spatial form was socially 'planned', in itself directly socially caused, that way. But there is also an element of 'chaos' which is intrinsic to the spatial. For although the location of each (or a set) of a number of phenomena may be directly caused (we know why X is here and Y is there), the spatial positioning of one in relation to the other (X's location in relation to Y) may not be directly caused. Such relative locations are produced out of the independent operation of separate determinations. They are in that sense 'unintended consequences'. Thus, the chaos of the spatial results from the happenstance juxtapositions, the accidental separations, the often paradoxical nature of the spatial arrangements which result from the operation of all these causalities. Both Mike Davis and Ed Soja, for instance, point to the paradoxical mixtures, the unexpected land uses side by side, within Los Angeles. Thus, the relation between social relations and spatiality may vary between that of a fairly coherent system (where social and spatial form are mutually determinant) and that where the particular spatial form is not directly socially caused at all.

This has a number of significant implications. To begin with, it takes further the debate with Ernesto Laclau. For in this conceptualization, space is essentially disrupted. It is, indeed, 'dislocated' and necessarily so. The simultaneity of space as defined here in no way implies the internally coherent closed system of causality which is dubbed 'spatial' in his *Reflections*. There is no way that 'spatiality' in this sense 'means coexistence within a structure that establishes the positive nature of all its terms' (Laclau 1990: 69). The spatial, in fact, precisely *cannot* be so. And this means, in turn, that the spatial too is open to politics.

But, further, neither does this view of space accord with Fredric Jameson's which, at first sight, might seem to be the opposite of Laclau's. In Jameson's view the spatial does indeed, as we have seen, have a lot to do with the chaotic. While for Laclau spatial discourses are the attempt to represent (to pin down the essentially unmappable), for Jameson the spatial is precisely unrepresentable – which is why he calls for an exercise in 'mapping' (though he acknowledges the procedure will be far more complex than cartography as we have known it so far). In this sense, Laclau and Jameson, both of whom use the terms space/spatiality, etc. with great frequency, and for both of whom the concepts perform an important function in their overall schemata, have diametrically opposed interpretations of what the terms actually mean. Yet for both of them their concepts of spatiality work against politics. While for Laclau it is the essential orderliness of the spatial (as he defines it) which means the death of history and politics, for Jameson it is the chaos (precisely, the dislocation) of (his definition of) the spatial which apparently causes him to panic, and to call for a map.

So this difference between the two authors does not imply that, since the view of the spatial proposed here is in disagreement with that of Laclau, it concords with that of Jameson. Jameson's view is in fact equally problematical for politics, although in a different way. Jameson labels as 'space' what he sees as unrepresentable (thus the 'crisis of representation' and the 'increasing spatialization' are to him inextricably associated elements of postmodern society). In this, he perhaps unknowingly recalls an old debate within geography which goes by the name of 'the problem of geographical description' (Darby 1962). Thus, 30 years ago H. C. Darby, an eminent figure in the geography of his day, ruminated that 'A series of geographical facts is much more difficult to present than a sequence of historical facts. Events follow one another in time in an inherently dramatic fashion that makes juxtaposition in time easier to convey through the written word than juxtaposition in space. Geographical description is inevitably more difficult to achieve successfully than is historical narrative' (ibid.: 2). Such a view, however, depends on the notion that the difficulty of geographical description (as opposed to temporal story-telling) arises in part because in space you can go off in any direction and in part because in space things which are next to each other are not necessarily connected. However, not only does this reduce space to unrepresentable chaos, it is also extremely problematical in what it implies for the notion of *time*. And this would seem on occasions to be the case for Jameson too. For, while space is posed as the unrepresentable, time is thereby, at least implicitly and at those moments, *counterposed* as the comforting security of a story it is possible to tell. This, of course, clearly reflects a notion of the difference between time and space in which time has a coherence and logic to its telling, while space does not. It is the view of time which Jameson might, according to some of his writings, like to see restored; time/History in the form of the Grand Narrative.[11]

However, this is also a view of temporality, as sequential coherence, which has come in for much questioning. The historical in fact can pose similar problems of representation to the geographical. Moreover, and ironically, it is precisely this view of history which Laclau would term spatial

> with inexorable logic it then follows that there can be no dislocation possible in this process. If everything that happens can be explained *internally* to this world, nothing can be a mere event (which entails a radical temporality, as we have seen) and everything acquires an absolute intelligibility within the grandiose scheme of a pure spatiality. This is the Hegelian–Marxist moment.
>
> (Laclau 1990: 75)

Further still, what is crucially wrong with both these views is that they are simply opposing space and time. For both Laclau and Jameson, time and space are causal closure/representability on the one hand and unrepresentability on the other. They simply differ as to which is which! What unites them, and what I argue should be questioned, is the very counterposition in this way of space and time. It is a counterposition which makes it difficult to think the social in terms of the real multiplicities of space-time. This is an argument which is being made forcefully in debates over cultural identity: 'ethnic identity and difference are socially produced in the here and now, not archeologically salvaged from the disappearing past' (Smith 1992); and Homi Bhabha enquires:

> Can I just clarify that what to me is problematic about the understanding of the 'fundamentalist' position in the Rushdie case is that it is *represented* as archaic, almost medieval. It may sound very strange to us, it may sound absolutely absurd to some people, but the point is that the demands over *The Satanic Verses* are being made *now*, out of a particular political state that is functioning very much in our time.[12]

Those who focus on what they see as the terrifying simultaneity of today, would presumably find such a view of the world problematical, and would long for such 'ethnic identities' and 'fundamentalisms' to be (re)placed in the past so that one story of progression between differences, rather than an account of the production of a number of different differences at one moment in time, could be told. That this cannot be done is the real meaning of the contrast between thinking in terms of three dimensions plus one, and recognizing fully the inextricability of the four dimensions together. What used to be thought of as 'the problem of geographical description' is actually the more general difficulty of dealing with a world which is 4-D.

But all this leads to a fourth characteristic of an alternative view of space, as part of space-time. For precisely that element of the chaotic, or dislocated, which is intrinsic to the spatial has effects on the social phenomena that constitute it. Spatial form as 'outcome' (the happenstance juxtapositions and so forth) has emergent powers which can have effects on subsequent events. Spatial form can alter the future course of the very histories which have produced it. In relation to Laclau what this means, ironically, is that one of the sources of the dislocation, on the existence of which he (in my view correctly) insists, is precisely the spatial. The spatial (in my terms) is precisely one of the sources of the temporal (in his terms). In relation to Jameson the (at least partial) chaos of the spatial (which he recognizes) is precisely one of the reasons why the temporal is not, and cannot be, so tidy and monolithic a tale as he might wish. One way of thinking about all this is to say that the spatial is integral to the production of history, and thus to the possibility of politics, just as the temporal is to geography. Another way is to insist on the inseparability of time and space, on their joint constitution through the interrelations between phenomena; on the necessity of thinking in terms of space-time.

NOTES

1 Acknowledgements to *New Left Review* 196: 65–84, in which this chapter first appeared.
2 Thanks to Ernesto Laclau for many long discussions during the writing of this article.
3 See, for instance, the discussion in Rustin (1987).
4 And in this sense, of course, it could be said that Laclau's space is 'political' because any representation is political. But this is the case only in the sense that *different* spaces, different 'cognitive mappings', to borrow Jameson's terminology, can express different political stances. It still leaves each space – and thus the concept of space – as characterized by closure and immobility, as containing no sense of the open, creative possibilities for political action/effectivity. Space is the realm of the discourse of closure, of the fixation of meaning.
5 See, for instance, Massey (1984); Gregory and Urry (1985); and Soja (1989).
6 It should be noted that the argument that 'the spatial' is particularly important in the current era is a different one from the one being made here. The argument about the nature of postmodernity is an empirical one about the characteristics of these times. The argument developed within geography was an in-principle position concerning the nature of explanation, and the role of the spatial within this.
7 See, for instance, Flax (1983); Harding and Hintikka (1983); Lange (1983); Flax (1990) and Hartsock (1990).

8 See, for instance, Dinnerstein (1987); le Dœuff (1991); and Lloyd (1984).

9 An alternative explanation of why such structures are labelled spatial is available. Moreover it is an explanation which relates also to the much wider question (although in fact it is rarely questioned) of why structuralist thought, or certain forms of it, has so often been dubbed spatial. This is that, since such structures are seen to be non-dynamic systems, they are argued to be non-temporal. They are static, and thus lacking in a time dimension. So, by a knee-jerk response they are called spatial. Similarly with the distinction between diachrony and synchrony. Because the former is sometimes seen as temporal, its 'opposite' is automatically characterized as spatial (although in fact not by Laclau, for whom certain forms of diachrony may also be 'spatial' (Laclau 1990: 42)). This, however, returns us to the critique of a conceptualization of space simply and only in terms of a lack of temporality. A-temporality is not a sufficient, or satisfactory, definition of the spatial. Things can be static without being spatial – the assumption, noted earlier, that anything lacking a transformative dynamic is spatial cannot be maintained in positive terms; it is simply the (unsustainable) result of associating transformation solely with time. Moreover, while a particular synchrony (synchronic form) may have spatial characteristics, in its extension and configuration, that does not mean that it is a sufficient definition of space/spatial itself.

10 However, the social sciences deal with physical space too. All material phenomena, including social phenomena, are spatial. Any definition of space must include reference to its characteristics of extension, exclusivity, juxtaposition, and so on. Moreover, not only do the relationships between these phenomena create/define space-time but also the spacing (and timing) of phenomena enables and constrains the relationships themselves. Thus, it *is* necessary for social science to be at least consistent with concepts of physical space, although a social-science concept could also have additional features. The implications for the analysis of 'natural' space – of physical geography – are similar. Indeed, as Laclau argues, even physical space is temporal and therefore in his own lexicon not spatial: 'the real – including physical space – is in the ultimate instance temporal' (Laclau 1990: 41–2). While I disagree with the labelling as spatial and temporal I agree with the sense of this – but why only 'in the ultimate instance'?!

11 I am hesitant here in interpreting Jameson because, inevitably, his position has developed over the course of his work. I am sure that he would not in fact see narrative as unproblematic. Yet the counterposition of it to his concept of spatiality, and the way in which he formulates that concept, does lead, in those parts of his argument, to that impression being given.

12 In 'Interview with Homi Bhabha' in Rutherford, J. (1990: 215). At this point, as at a number of others, the argument here links up with the discussion by Peter Osborne in his 'Modernity is a qualitative, not a chronological, category'.

REFERENCES

Anderson, P. (1991) 'A culture in contraflow', *New Left Review* 180: 41–78; 182: 85–137.

Bhabha, H. (1990) 'The third space: interview with Homi Bhabha' in J. Rutherford (ed.) *Identity: Community, Culture, Difference*, London: Lawrence & Wishart, 207–21.

Darby, H. C. (1962) 'The problem of geographical description', *Transactions of the Institute of British Geographers* 30: 1–14.

de Beauvoir, S. (1972) *The Second Sex* (1st edn 1949), trans. H. M. Parshley, Harmondsworth: Penguin.

Dinnerstein, D. (1987) *The Rocking of the Cradle and the Ruling of the World*, London: The Women's Press.

Flax, J. (1983) 'Political philosophy and the patriarchal unconscious: a psychoanalytic perspective on epistemology and metaphysics', in S. Harding and M. B. Hintikka (eds) *Discovering Reality: Feminist Perspectives on Epistemology, Metaphysics, Methodology and Philosophy of Science*, Dordrecht: D. Reidel Publishing Company, 245–81.

—— (1990) 'Postmodernism and gender relations in feminist theory', in L. J. Nicholson (ed.) *Feminism/Postmodernism*, London: Routledge, 39–62.

Foucault, M. (1986) 'Of other spaces', *Diacritics* 16: 22–7.

Gregory, D. (1990) 'Chinatown, part three? Soja and the missing spaces of social theory', *Strategies: a Journal of Theory, Culture and Politics* 3.

Gregory, D. and Urry, J. (eds) (1985) *Social Relations and Spatial Structures*, Basingstoke: Macmillan.

Harding, S. and Hintikka, M. B. (1983) 'Introduction', in S. Harding and M. B. Hintikka (eds) *Discovering Reality: Feminist Perspectives on Epistemology, Metaphysics, Methodology and Philosophy of Science*, Dordrecht: D. Reidel Publishing Company, ix–xix.

Hartsock, N. (1990) 'Foucault on power: a theory for women?' in L. J. Nicholson (ed.) *Feminism/Postmodernism*, London: Routledge, 157–75.

Jameson, F. (1991) *Postmodernism, or, the Cultural Logic of Late Capitalism*, London: Verso.

Jay, N. (1981) 'Gender and dichotomy', *Feminist Studies* 7(1): 38–56.

Laclau, E. (1990) *New Reflections on the Revolution of our Time*, London: Verso.

Lange, L. (1983) 'Woman is not a rational animal: on Aristotle's biology of reproduction', in S. Harding and M. B. Hintikka (eds) *Discovering Reality: Feminist Perspectives on Epistemology, Metaphysics, Methodology, and Philosophy of Science*, Dordrecht: D. Reidel Publishing Company, 1–15.

le Dœuff, M. (1991) *Hipparchia's Choice: an Essay Concerning Women, Philosophy, etc.*, Oxford: Blackwell.

Lefebvre, H. (1991) *The Production of Space* (1st edn 1974), Oxford: Basil Blackwell.

Lloyd, G. (1984) *The Man of Reason: 'Male' and 'Female' in Western Philosophy*, London: Methuen.

Massey, D. (1984) *Spatial Divisions of Labour: Social Structures and the Geography of Production*, Basingstoke: Macmillan.

—— (1993) 'Power-geometry and a progressive sense of place', in J. Bird *et al.* (eds) *Mapping the Futures*, London: Routledge.

Osborne, P. (1992) 'Modernity is a qualitative, not a chronological, category', *New Left Review* 192: 65–84.

Ross, K. (1988) *The Emergence of Social Space: Rimbaud and the Paris Commune*, Basingstoke: Macmillan.

Rustin, M. (1987) 'Place and time in socialist theory', *Radical Philosophy* 47: 30–6.

Rutherford, J. (ed.) (1990) *Identity: Community, Culture, Difference*, London: Lawrence & Wishart, 207–21, 215.

Smith, M. P. (1992) 'Postmodernism, urban ethnography, and the new social space of ethnic identity', *Theory and Society*.

Soja, E. (1989) *Postmodern Geographies: the Reassertion of Space in Critical Social Theory*, London: Verso.

Stannard, R. (1989) *Grounds for Reasonable Belief*, Edinburgh: Scottish Academic Press.

Wilson, E. (1991) *The Sphinx in the City: Urban Life, the Control of Disorder, and Women*, London: Virago Press.

9

BLACK TO FRONT AND BLACK AGAIN

Racialization through contested times and spaces

Barnor Hesse

> When I heard about Bristol and saw what happened in Liverpool, I said to myself, what a co-incidence. Two former slaving ports. The great-great-grandchildren have come here to take revenge.
>
> (Ludwig Hesse (quoted in Dennis 1988))

> All I can think of are places and scenes of all the places we've been.
>
> (Scott-Heron 1977)

> A sojourn, as I define it . . . implies not only a journey, but also a stay. It is much more than a simple journey from point A. to point B.
>
> (Zhana 1988)

> U.K. Blak letting you know that we're about
>
> (Wheeler 1990)

INTRODUCTION

The articulation of times and spaces mark the identification of politics and cultural identities as resistant to any form of totalizing theoretical practice. The point of this observation is one of its corollaries: time and space subvert theoretical practice. This is always and everywhere the problem that confounds narration, the 'infrastructure' (Gashe 1986) of theory, in its guise as the difference between a 'ground' (theorizable) and what is 'grounded' (theorized). If our objective is to illuminate the political, its institution or modalities, inevitably we encounter experiences and events from 'other' narratives, from elsewhere or another time. But these 'outsides' can only enter 'our' narrative if we remember them in time or can locate a relevant space. And of course it is the narrative, more significant as product than production, which recruits events and occurrences as sequences in the provision of a commentary for comment, a discursive universe in which analyses find a natural orbit. Foucault (1984: 14) has described the insinuative manœuvres of the ubiquitous narrative: 'There is scarcely a society without its major narratives which are recounted, repeated and varied; formulae texts, and ritualized sets of discourses which are recited in well defined circumstances.' The history of Black settlements in Britain since 1945, whether considered in its political or cultural impresses, seems now to be the preserve of a formulaic, variable narration in which the temporal and spatial inscriptions of difference in Black settlement and their political implications are simply repressed.

The dominant and popular political narrative of 'race' in Britain almost always begins, opens, within the framework of post-1945 Britain, this period of postwar immigration is referred to as the period of settlement, as if Black settlement began in the second half of the twentieth century. In this narrative, '1948' stands as a nodal point in any subsequent periodization. The arrival of the *Empire Windrush* on 22 June 1948 with 492 Jamaicans on board (Fryer 1984: 372) was a national media event. These journey men and women symbolize in the British popular imagination various themes: the 'coloured' trickle that became a black flood, the encounter with the 'motherland', the arrival of coloured neighbours or workers, the subsequent discovery of the 'colour bar'. What '1948' fails to signify, interestingly enough in this recuperation of settlement chronology, that is within the narrative, are the 'racist riots' which erupted in August in Liverpool where Black settlement was so to speak, already in place. According to Fryer (ibid.: 307) by this time there were about 8,000 Black people in Liverpool. Certainly by 1953, the numbers of 'West Indian workers entering Britain', had not exceeded this figure. Yet significant though this event was regionally, it somehow manages to evade inclusion in the national political narrative of 'race' as settlement, even in Fryer's own account despite the attention he devotes to it. The usual periodization of postwar settlement (1948–58) from Windrush to Nottinghill and Nottingham simply leaves this out. For example, Stuart Hall has written: 'the first signs of an open and emergent racism of a specifically indigenous type appears, of course, in the race riots of Nottinghill and Nottingham in 1958' (Hall 1978: 37). Now it is arguable that this simply 'forgets' not just earlier settlements, but settlements elsewhere; that is to say, by 1958, Liverpool is not simply the region of an earlier (temporalized) settlement, it is a settlement elsewhere (spatialized). The point is even if it is agreed that Black historical research is of recent origin, the point is where it exists it does not manage to disturb the narrative in place.

The historical research of Fryer (ibid.: 375) does not disrupt this. The desire to produce a national narrative shows itself incongruently in observations which deregionalize the experience of Black settlement. For example, he writes 'to the prejudiced and ignorant majority it seemed "strange and out of place" merely to see black people in the centres of large industrial centres in Britain,' (ibid.). This could hardly be a description of Liverpool or Cardiff, where strangeness had already been overcome by the familiarity of urban-bred racial antagonisms. Once again, despite the historical evidence to the contrary, the narrative form of the postwar settlement experience is in place, paradigmatically. Fryer characterizes the ten-year period 1948–58 as the 'laissez faire period' in government immigration control. This is a regular periodization in the narrative, it marks the retrospective run-up to the race riots of 1958 in Nottinghill and Nottingham.

In considering this, it is worth thinking laterally against the grain of this narrative, since the politics of journeying in Black Britain has a number of directions which resist any neat codification in a historical narrative that presents an uneven regional configuration as a linear sequence. What is required, as Berger argues in the context of the novel, is a 'change in the mode of narration' (Berger, quoted in Soja 1989). In analysing the social movement of Black politics in Britain, 'anomalous' cities like Liverpool (see Small 1991) suggest, 'it is scarcely any longer possible to tell a straight story sequentially unfolding in time' (Berger, quoted in Soja 1989). The point is that Black experiences slip effortlessly out of our grasp unless we are aware of 'what is continually traversing the story laterally' (ibid.). Berger's comments are useful in trying to think the contingent configurations of Black journeying, the recurrent and various traversing of time and space. This historical narrative needs to be deconstructed not only in its attachment to an institution in a linear form, it needs to leave open the impossibility of closure.

On the face of it, the title of Ron Ramdin's book (1987) *The Making of the Black Working Class in Britain*, seems to gesture in a lateral direction. While presenting a chronology from 1555 to the 1980s, Ramdin's approach to racial thematics gives some sense of the development of political cultures in terms of the 'earlier and the elsewhere'. It is interesting to note that Ramdin's characterization of postwar settlement seems aware of the radical discrepancies incurred in the narrative by restrictive time-spatial boundaries, and he attempts a resolution. He observes that the newcomers who arrived in Britain between 1947 and 1962 dispersed to several parts of the country. Although the majority stayed in London, many went to Liverpool, Manchester, Cardiff, Birmingham, Bristol, Nottingham, Bradford, Leeds, Leicester, Wolverhampton and other towns and cities (ibid.: 187). He then announces, in line with the dominant narrative tradition, that his discussions of postwar immigration will 'concentrate essentially on the problems West Indians faced in Nottingham and Nottinghill in London'. The rationale for this focus, and perhaps the direction of the historical narrative, is provided in curious terms by Ramdin (ibid.: 758). He suggests: 'in terms of both geography and time, the focus of British racism had shifted from Cardiff and Liverpool to Nottingham and Nottinghill'. But *whose* focus we might ask; and according to whom had this shift occurred? There is, however, a sense in which the *historiography* of the period does shift to London as a signifier of national concern. For example, Solomos has argued:

> The 1958 race-riots in Nottingham and Nottinghill are commonly seen as an important watershed in the development of racialised politics in Britain. It is certainly true that the events in these two localities helped to

> bring to national prominence issues which had previously been discussed either locally or within government.
>
> (Solomos 1989: 48)

In many respects, what marks the significance of this postwar period is not immigration and new settlement as such, but rather the terms upon which settlement was now being fought, with clearer expectations and demands being made in the light of the urgency of anti-colonial struggles 'back-a-yard' and 'back home' and perhaps more importantly in this light, of Black men having fought in the Second World War. The convergence of unsettled, racial settlement patterns is what Britain experiences acutely in this period. As Solomos (ibid.: 30–6) has noted, by the end of the nineteenth century, Black seamen were settled populations in Liverpool, London, Cardiff, Bristol and the port-towns: it was the 'politics of black migrants and settlement in Britain in the period of the early twentieth century' which was the impress upon which the 'terms of political debate and domestic ideologies and politics towards coloured workers and the communities were formed in embryonic fashion'. What we need to question here is whether there is a social logic in the experiences of Black settlement in Britain and how that might be theorized. However, it is with a wider and deeper notion of 'settlement' that we can begin to think about the logic of a Black social movement, as a poetics of journeying and a politics of reinscribing times and spaces. As I shall argue by example in the rest of this chapter, a critical deconstruction of a conventionally 'British' race/nation narration turns various textual expectations and assumptions if not upside down then back to front.

HETEROGRAPHIES OF SETTLEMENT

The politics of 'race' in general and Black politics in particular does not appear to have a theoretically defined status in the literature which constitutes the field of 'radical' social and political thought in Britain. This virtual theoretical silence is compounded by the fact that between the various iterations of the signifier 'Black-political' and the contingent, diasporic contexts of its signified, there is a sense in which to speak of a Black social movement in Britain at all requires that we reconstruct a language which makes this possible. This raises another level of complexity in the theorization of Black politics, its indissolubly anti-Western and Western 'origins'. Without being too categoric, this appears to the rationale of Robinson's argument where he writes:

> The social cauldron of Black radicalism is Western society. Western society, however, has been its location and its objective condition but not – except in a perverse fashion – its specific inspiration. Black radicalism is a negation of Western civilization, but not in the direct sense of a simple negation.
>
> (Robinson 1983: 96)

Developing this further requires that we see the modern settlement of Black communities in the West as problematizing hegemonic designations of time and space. What might be termed the 'racialization of modernity' was fundamentally disruptive and productive of times and spaces as politics. On this account, the modern period had a Columbian opening in '1492' (Todorov 1982) and the so-called Age of Discovery was contemporaneous with Amerindian ethnocide and African enslavement in the age of 'white supremacist logics' (West 1988) and the configuration of the world as a 'globality'. This sets up Manning Marable's important observation: 'Politics for those in the historical crucible of exploitation and destruction becomes both a means of human affirmation and a mode of resistance to the excruciating apparatus of systemic violence in which people of the African diaspora find themselves' (Marable 1985: 3).

Modernity continues to project a nightmarish shadow over the formations of Black cultural and political identities. The 'here and now' of our 'social imaginaries' (Laclau 1990) encounter articulations of timed-spaced tracings of 300 years of African enslavement; the exploitative, violent division of the world into sublime Europe and savage/exotic non-Europe (see Fanon 1963; Amin 1988; Said 1978; Mudimbe 1988); and the alternating invocations of 'peoples without histories' (Wolf 1982) and 'natives' from another time (Fabian 1983). There are many levels and dimensions in Black critical thinking, but what unites them as a 'regularity in dispersion' (Foucault 1972; Laclau and Mouffe 1985) is a mundane 'incredulity' towards Eurocentrism as the Grand Narrative (cf. Lyotard 1984), together with an abiding problematization of 'the way we think about and experience time and space, history and geography, sequence and simultaneity, event and locality, the immediate period and region in which we live. Modernity is thus comprised of both context and conjecture' (Soja 1989: 25). The Black experiences of slavery, colonialism and contemporary European racism form part of this context and conjuncture; this provides a theoretical rationale for developing an understanding of the discursivity of politics and culture in the spatialized and temporalized Black presence in twentieth-century Britain and elsewhere.

Under the sign of Black cultural theory there is a familiar trope of recurrent, uneven movement and residence, of journeying beyond (dis)placement and (re)location, which is inscribed in various Black enactments of cultural autobiographical expression: writing, signifying, testifying, reasoning, organizing, demonstrating and campaigning illustrate multiple ways in which the poetics in figuring and remembering Black experiences merge with day-to-day racialized existences. These employ diverse forms of what Toni Morrison (1990) describes as a 'kind of literary archaeology' which uses 'the basis of some information and a little bit of guesswork [in] a journey to a site to see what remains were left behind and to reconstruct the world that these remains imply'. In the poetics of this journeying, the pasts and futures of Africa, the Caribbean Americas and Europe are congealed and decentred in various

Black modes of representation. Yet what particularly disturbs the semblance of any contemporary settlement in Black cultural identities are diverse struggles over forms of axiological restitution and communal edification in the institution of the 'social'.

What I describe here as axiological restitution relates to the profound use of memory and popular historiography in Black communities as a basis for socially legislating the recovery from distortion or mutilation the ethics and aesthetics of Black bodily and cultural integrity in modernity. This can combine, although not necessarily or predictably, with the critical desire for 'communal edification' in which the different immediacy of economic inequalities, social injustices and civil representation concentrate the heart and soul of survival tactics and focus objectives for political transformation. It is important to conceive these forms of struggle as operative on a lower, more mundane register in the confrontation with a racialized capitalism (Robinson 1983) than a systematically coded critique might suggest. These ideas bear some affinity with and are perhaps presupposed by what Paul Gilroy (1991: 11) describes as the 'politics of fulfilment' in Black musical culture which espouses the idea that a future society will be able to realize unaccomplished social and political promises. However, they also elaborate Gilroy's related discussion of the 'politics of transfiguration', which 'emphasizes the emergence of qualitatively new desires, social relations and modes of association within the racial community of interpretation and resistance and between that group and its erstwhile oppressors'. In expanding, fracturing and compressing the inscriptions of Black communities' 'comings and goings' the interactivity of these struggles for politics enunciate in their global and local institutions what I describe as 'heterographies of settlement'. Some idea of my thinking on this is available through a pluralized interpretation of Gayatri Spivak's reflections on the staging of autobiographical representations: 'Experience is the staging of experience. One can only offer scrupulous and plausible accounts of the agencies or mechanics of staging. . . . 'What is it to stage?' 'What is to be staged?' (Spivak 1992: 9).

Theoretically, this prefigures a politics of journeying in which time and space are both thought and fought. It is the staging which has to be settled: every stage and staging is a story and 'other stories' (see Aareen 1989). In terms of these heterographies, what is it to be settled? How can we think the settlement of Black communities in Britain during the twentieth century as a social movement or a politics? The answer I theorize circuitously in this chapter can also be put succinctly as a proposition: the settlement is the movement. By this I am arguing that the development of Black identities or of self-help organizations, the visibility and audibility of urban Black styles and sounds, the campaigns against police racism, the maintenance of homes and families and the struggles in and for jobs and education are all embroiled in the same 'racial' antagonisms (Gilroy 1987), which reveal the equivocations and ambivalences in the conditions for settling. Theorizing in this manner necessitates a clean break with a narrow temporalized notion of settlement, as if it were a discrete moment or an ephemeral legacy of immigration, disappearing generationally like a 'foreign' accent.

In a detailed discussion of the configuration of 'race' in social movement theory and the urban environments of Britain, Gilroy (1987) raises many of the issues that are symptomatic of this discussion, but fails to theorize these sufficiently, leaving unexplored much of the answer to the question, 'what makes Black social movements move'? Here we have to return to the problematization of settlement and consider it in theoretical terms. Gilroy (ibid.: 223) comes close to this where he argues that '[a]nalysis of black Britain must be able to address the synchronic, structural aspects of the movement as well as its diachronic, historical dimension'. The problem here lies in conceiving the notion of movement, without suggesting a strategic logic in which 'conflicts over the production of urban meanings' (ibid.: 228) or the displacement of the language of class and 'race' by the language of community in the 'political activity of black Britain' (ibid.: 230) necessarily, has a forward march. The latter would imply a political subject like a party or some other organization, or that there is a centre of direction in the movement. The problem of movement is also the problem of understanding the politics of settlement which, like the impact of the twentieth-century Black presence in Britain, is in effect a 'strategy without strategists' (Dreyfus and Rainbow 1986). I believe what confounds much analysis is how to think settlement as the reiteration of Black community when 'the logic is perfectly clear, the aims decipherable, and yet it is often the case that no one is there to have invented them, and few who can be said to have formulated them' (Foucault, quoted in Dreyfus and Rainbow 1986: 187). Moreover, the regionalization (decentring) of Black people in Britain, as well as our dialogical, proximal relation to the racialized experiences of Asian communities, suggests any configuration of heterographies should be handled contingently. Nevertheless I think there are at least three layers to Black settlement which can be provisionally unearthed here. First, it is a dynamic and uneven social process, it is not something that simply descends on Britain like a plane or an overlay of snow, it is not a 'swamping'; it is both concentrated and threadbare in its deepening of the complexity of the social and extending the ethnicization of its horizons. It is therefore ethno-spatially dispersed. Second, settlement carries with it all the senses of coming to terms with conditions, terms, experiences, set-backs and promises; it is about settling what is contestable and what is acceptable, settling down and settling up. It is, therefore, also an ethno-temporal articulation. The third layer is the 'reactivation' (Laclau 1990) of the previous two forms of sedimentation, it reveals discordant series of movements in the configuration of community as settlement or as settled, it is a (re)interrogation of time and space.

Although not often expressed in these terms, this politics of time and space has a complex cultural heritage in Black discourses. Despite the suppression of its intonation, it is this which provides the basis for the diasporic status of 'Black' as a 'floating signifier' in the world post-1492. How else can we understand the chilling realization that 'at the heart of Black politics is a series of crimes' (Marable 1985: 1)? A necessarily aporiatic sense of what I am suggesting here can be seen in Edouard Glissant's (1989) significant reiteration of 'the quarrel with history'. He reasserts a memory of the Caribbean as a 'site of history characterized by ruptures . . . that begin with a brutal dislocation, the slave trade'. This produced a split in a dominant code of temporality, the radical progressivism of Western historicity. Thus, argues Glissant (1989: 62), 'our historical consciousness could not be deposited gradually and continuously like sediment'. This meant there was no actual root for a comparable European 'totalitarian philosophy of history', instead there was only a potential re-routing in a 'context of shock, contraction, painful negation and explosive forces'. What Glissant describes as the 'dislocation of the [historical] continuum' symptomatically marks the emergence of the African diaspora (Harris 1982; Thompson 1987) and the 'Black Atlantic world' (Gilroy, forthcoming). These categories can be described as 'social imaginaries' (Laclau 1990: 64), representations of 'fullness' in a projection or an identity, located beyond the precariousness and dislocations characteristic of the everyday world. The diasporic imaginary attempts to synthesize the radical dispersal, displacement and contiguity of populations invested with an African genealogy in the post-1492 world, descending down a line of resistances and affirmations in encounters with contemporary white racism, nineteenth-century colonialism and sixteenth- to nineteenth-century enslavement. Recognizing the African diaspora as a social imaginary takes us beyond the empiricist 'binary formation – us and the others, a residual construction surviving from the master/slave heritage' (Clark 1991: 42), so characteristic of extreme forms of African-American particularism and some versions of Afrocentrism (cf. Asante 1987; 1988). Laclau puts the point effectively when he writes, 'The imaginary is a horizon: it is not one among other objects but an absolute limit which structures a field of intelligibility and is thus the condition of possibility for the emergence of any object' (Laclau 1990: 64).

Returning to Glissant (1989), I want to argue that the diasporic imaginary itself is vitally emergent in the conditions set by the 'inability of collective consciousness to absorb' the enormity of the traumas associated with slavery and also the ultimate impossibility of a collective consciousness. Black diasporic imaginaries therefore inscribe a poetics of time and space which is always unimaginable. paradoxically, Glissant's suggestion that the dislocation of the historical continuum produces a 'non-history' complete with the erasing of the collective memory, can be taken as radically productive. It facilitates focus on a system of organizing experiences, space, which is the project of another temporality, not only a different history, but in effect the history of different spaces: the African diaspora.

The racial dislocations of modernity (not only slavery but colonialism) generate what I want to call the Black temporality of the 'cut' (see Snead 1984); a sense of time or timing which is played out, endured and occasioned through discontinuous forms of repetition. James Snead introduced the concept of the 'cut' in a discussion of Black musical forms. The 'cut' is a dissipated beat, a return to repetition, in which it continually 'cuts' back to the start. What this also encapsulates is a reflexive temporality where the cut is both the site of dislocation and the process of collecting memories to return to it. Beyond this then the 'cut' represents the traumas and inspiration condensed in the Black diasporic condition, signifying 'an abrupt, seemingly unmotivated break . . . with a series already in progress and a willed return to a prior series' (my emphasis). The familiar meditative and archaeological comprehension of African dislocations thrown up here combines with the resonating pains of 'origins' as the cutting edges of this temporality. Within and against the evolutionary mystic of Western historicity, 'Black expressive cultures' (Gilroy 1987) are impaled on racist logics both near and far, irreducible to the separability of time and space. It is worth noting that this cut of temporality is also the cult of memory in Black popular culture, especially music. For example, during the 1970s the Jamaican reggae artist Burning Spear recorded one of many invocations, heard in Britain's Black communities, to 'remember the days of slavery' (Burning Spear 1975) and Black echoes of this could still be heard in the 1980s when, among others, Steel Pulse, the reggae band once based in Handsworth (Birmingham), 'cursed that day they made us slaves' (Steel Pulse 1982).

All this makes the designation of the present difficult: it is always imperfect. The Black temporality of the cut 'builds accidents into its coverage, almost as if to control their unpredictability' (Snead 1984). Living in Britain for Black communities engaged in the modes of settlement I have outlined, is a return journey through various uncertain racisms where the point is not simply to interpret but to transform the settlement experience. The effect is to 'confront accident and rupture not by covering over but by making room for them inside the system itself' (ibid.). Black community settlements are further temporalized in this way by the instabilities (tensions and dimensions) between and inside the various geographical signifiers of space: place, neighbourhood, community, environment, countryside, nation. As discursive locations of heterographies, these are racialized sites of 'other-presencing' and disjunctive presences. If this 'signifying difference' (Gates 1984) arises from 'two contexts, two traditions – the western and the black', then perhaps, Bhabha's (1990: 6) suggestion of a decentred form of contextualization gives a directionality and staging to Black cultures of politics. He argues that is it only when the 'western nation' is perceived as 'one of the dark corners of the earth that we can begin to explore new places from which to write histories of peoples and construct theories of narration'. The heterographies of Black settlement represent a trenchant critique of the historicist reification of the British nation as presence. In other words the presence as such is not a unity but a 'singular plural which no simple origin will ever have preceded' (Derrida in Clark 1987: 133). If anything captures effectively the mood of Derrida's critique of the metaphysics of presence, it is surely the racialized fusion of time and space in 'official' and 'civil' projections of the heritage and nature of the British nation (see Wright 1985). Its national, geographic past is established through 'relations of equivalence' (Laclau and Mouffe 1985) between the 'recent experience of economic and imperial decline . . . persistence of imperialist forms of self-understanding . . . early depopulation and redevelopment of settled communities' (Wright 1985: 25). What arranges these items so that they cease to be merely an arbitrary inventory of the past is their unifying and 'othering' effects. Their ritual, national invocation distils a prescribed authentic sense of a shared past and emplacement. The British nation is both people and terrain, an almost timeless, exclusive, static 'ethnoscape' (see Appadurai 1990; Hesse *et al.* 1992). The 'here and now' is racially articulated as very nearly present, devoid of fundamental contingency or difference and insistent that 'in spite of our always fragmenting experience, somewhere there must exist a redeeming and justifying wholeness' (Hawkes 1977: 146). In the postwar period, it is not only the project to put at least a lower-case '(g)reat' back into Britain, which encodes a dominant 'white' representation of the nation, but the intensity of response and agitations which surrounded yet another extraction of 'racial discourse from the body politic'. During the twentieth century, it might be said, 'race' occurs three times. The first time as 'coloured colonials'; the second as 'coloured immigrants'; and the third as Black (and Asian) citizens, each dimension marked by descendants who elided, reversed and lateralized this sequence. Succinctly, it might be argued that racism in prewar Britain was largely anchored in the effluence of the politics of colonialism and the expansion and welfare of the 'coloured quarters' (Little 1948) in the nation's ports; after the war this both continued and changed. With the catalyst of the economic migration from the departing colonies, a racial politics of British nationalism and ethnic density begins to occupy the nation's regionalization of itself as an imagined community (Anderson 1983).

RACIAL ANTAGONISMS AND DISLOCATIONS

In different regions of Britain, the primary basis for the analysis of racial antagonisms is the identification of the forms through which 'racial meanings . . . provide the basis for action' (Gilroy 1987: 27). I have argued elsewhere that the linkage between 'territoriality, the assertion of imperial white identity and racial harassment [arises] among various individuals and groups in white communities who regard themselves in racial or cultural terms to be defending their space against change and transformation' (Hesse *et al.* 1992: 173). It is important at this point to consider what is meant theoretically by 'antagonism', in order to grasp how it structures the racialization of social relations. This requires we understand the notions of 'white society' or 'multicultural society' as discursive attempts to constitute the idea of a social totality as an objective and closed system of differences (Laclau and Mouffe 1985). An antagonism emerges through the logic of what Laclau and Mouffe, following Derrida, describe as a 'constitutive outside', this is the excluded relational identity (e.g. the non-whites) against which a privileged social identity is defined and distinguished. Antagonism describes the logic that leads to the dislocation of social identities: it 'blocks' their fullness. In terms of our discussion, however, the 'multicultural society' and white society are in fact related in a mutual racial antagonism – neither concept can be conceived outside a direct negative relation to the other. They constitute each other's significant constitutive outsides. The constitutive outside is inscribed as the 'non- or anti-essence that violates the boundary of positivity' by which the respective other has been 'formerly thought to be preserved in its as-such' (Staten 1984: 18).

Racial antagonisms in Britain are a contested experience of 'the limit of the social' (Laclau 1990) since the affirmation both of 'white society' and 'multicultural society', in violating each other's boundaries, also becomes 'the positive conditions of the possibility of the assertion of that boundary' (Staten 1984). There is a slippery paradox here that can be illuminated with reference to a popular rallying cry in the 1950s, which apostrophized 'Keep Britain White!' This still symbolizes the incidence of racial antagonism in the 'social' in so far as it reveals the dislocation of British identity being experienced as a crisis. But, the very conditions which made possible the assertion of British identity as white identity (i.e. Black and Asian settlement), were in fact the only conditions against which a white British identity could coherently emerge to define itself. Whatever the popularity of this interpellation, it conceded what it had to deny was possible even on its own terms, namely the existence of 'non-white Britain', its constitutive outside. Consequently, an exclusive white Britain is blocked even as a conception since imagining the latter depends on the very factors (i.e. the racialized others) which it sees as blocking it. Black and Asian people are crucial to the existence of 'white society'.

Understood spatially, racial antagonisms transform the enunciative positionality of British identity, 'multicultural' incarnations of the social symbolize its politicization against an exclusive ethnicization (i.e. racialization) by articulating 'Britain' to different chains of signification. Whether these are Black or Asian, British identity becomes inscribed in a process of decentring principally through forms of urban racial regionalization. The key to understanding this lies in rethinking the racial politics catalysed by the economic migration of Black people and others (from the former colonies) during and since the 1950s. Traditionally this experience has not been subject to 'detailed or comparative studies of the interplay between race and politics at the local level' (Solomos 1989: 86), the emphasis therefore has privileged the historical 'role of black and anti-racist political mobilisation in the context of British political life since 1946' (ibid.: 158). It leaves to one side the geographical consequences of settlement and 'race' in sustaining cultures of Black politics in Britain's regions and cities throughout the twentieth century. Here I can only offer some provisional comments on how this analysis might be initiated. It is possible to consider the racial harassment riots of 1919 in Cardiff and Liverpool, of 1948 in Liverpool, of 1958 in Nottinghill and Nottingham, and the urban Black uprisings against policing during the 1980s in Bristol, Brixton, Tottenham, Handsworth and Liverpool 8 as discontinuous nodal points which mark out not only alternative 'local' historiographies but reinscribe the cities as Black culture's geographic expressions.

National insides-outside

In order to develop this theoretically, I want to discuss another way of thinking the 'race', 'time' and 'space' in terms of a critique of Laclau's conception of dislocation (Laclau 1990: 41–59). Laclau argues there are three crucial dimensions to dislocation understood as revealing the general impossibility of a structure (e.g. the British nation) constituting its own self-sufficiency. The first is that 'dislocation is the very form of temporality'. For Laclau, spatiality always denotes a pre-given structure, a closed system of experiences which take the form of a repetition or a teleology. Any radical departures from this structure cannot be represented in terms of the structure (space). In so far as they disrupt it, it is an 'event' which points to the 'failure of hegemonization' to enclose all representation within the structure. Laclau describes this as temporality. The emergence of the previously unrepresentable is the dislocating factor. Now in one sense we can apply this to our discussion of Black settlement. The post-1945 'white' experience of immigrants was of a temporal intrusion, an event, which dislocated the apparently closed system of British national, landscape representations. One problem with this, however, is its suggestion of a singularity in the spatial and the discreteness of the event which precludes the coexistence of alternative or multiple representations and effaces the continuity of dislocation. However, let us consider Laclau's second dimension, namely that 'dislocation is the very form of possibility'. By this he does not mean a possibility immanent within a system as in a succession but rather the opening up of 'possibilities, of multiple and indeterminate re-articulations for those freed from its coercive force and who are consequently outside it'. In other words, 'the temporalization of spaces' widens the field of the possible. It is in this sense that I suggested the possibilities of 'white society' and 'multicultural society' emerged once the imperial codification of 'Great' Britain was dislocated. Yet this raises a further problem. Given the coexistence of these re-articulations, to what extent does this imply a historicist transition rather than a post-historicist (Vattimo 1988) displacement and thus render indeterminacy less contingent? Laclau's third dimension describes dislocation as 'the very form of freedom'. Once freedom is understood as 'the absence of determination', this means that no 'subject' is fully constituted – indeed, the 'subject' is a 'lack within the structure', it takes the form of a metaphor of an 'absent structure' or an 'absent fullness'. In this sense, the 'subject exists because of dislocation in the structure', therefore its freedom is not the freedom of a 'subject with a positive identity', but rather the freedom of a 'structural fault which can only construct identity through acts of identification'. The usefulness of this formulation lies in its application to the analysis of cultural identities. Where Laclau argues that the re-articulation of a dislocated structure is an 'eminently political re-articulation', this describes the logic of racial antagonism because 'efforts to re-articulate and reconstruct the structure also entail the constitution of the agent's identity and subjectivity'. This focuses the racialization of Britain. Its populations and landscapes, its histories and geographies are represented in the denouement of racial antagonisms which dislocate the identities of 'white territorialists' (Hesse *et al.* 1992) and Black and Asian settlers alike. The politics of 'race' becomes a politics of identification: 'To understand social reality, then, is not to understand what society is, but what prevents it from being' (Laclau 1990: 44).

There are, however, a number of problems that attach themselves to Laclau's use of time and space which need to be addressed before they can coherently be accommodated in this discussion. The explanatory power of Laclau's thesis turns on the acceptance of his proposition that 'temporality must be considered as the exact opposite of space' and that 'politics only exist in so far as the spatial eludes us' (ibid.: 68). For Laclau, 'space is the basis of representation', it is the 'objective' institution of the social, the realm of the political, while 'time' is the subversion of representation, the dislocation of the social's institution. The former affirms coexistences, the latter disrupts them. In attempting to use the insights of Laclau, it is important to recognize that it is only by contesting the particular figurations of concepts like time and space that we can geographically and historically contextualize those politics of difference which are engaged in and by the production, exchange, distribution and consumption of social imaginaries. It is along the variable differential axes of time and space, in the global and local politics of 'race', that the designation of context in this sense defines its undecidability. Through a deconstruction of Laclau's ideas we can begin to illuminate the argument theoretically. His separation of time and space is where we need to focus. It seems to imply that space, as an order of coexistences is analogous to 'Being' as presence (in the Heideggerian sense) and that time as the disorder of coexistences is therefore non-being or non-presence. In other words, it is the negativity of time which is opposed to the positivity of space. The problem here is how do we respond to Derrida's insistence that:

> Time is that which is thought on the basis of Being as presence and if something – which bears a relation to time, but is not time – is to be thought beyond the determination of being as presence, it cannot be a question of something that is called time.
>
> (Derrida 1982: 60)

This suggests that Laclau's conception of the 'temporal' is preceded by an undisclosed appeal to time, to a 'pre-comprehension of time, and within discourse, to the self-evidence and functioning of the verb's tense' (Derrida 1982: 50–1); this takes place inescapably within the horizon of time. This problem is compounded by the notion of dislocation itself, which as the very form of temporality (ibid.) is only distinguished as such from 'antagonism' which is prior to it and 'space' to which it is subsequent. What this suggests is that not only is dislocation disruptive of a spatial order, it is eruptive of a temporal order. If we further disturb the polarity of Laclau's time/space opposition, it is possible to argue, through reinscription of its terms, that the opposition in so far as it exists is not that between negativity and positivity, but between different orderings of 'order' and the 'social' (two negativities of presence). In effect, these are two disorders, one of dislocated coexistences, that is space, and the other of dislocated successions, that is time. This means recognizing the being of the 'social' as that which irreducibly 'fuses together evidential, spatial and above all temporal motifs' (Wood 1989: 3). Dislocation is therefore an undecidable form of time-spatial disordering. The departure for Black cultural and political analysis now becomes an exploration of the related racial logics of timed-spaced 'outsides' in an interrogation of the social. This should, of course, include establishing the conditions under which an 'outside' becomes racially antagonistic.

Diasporic outsides-inside

A different way of figuring and developing some of these issues can be pursued through a consideration of Alice Jardine's (1985) discussions of 'gynesis'. She argues that time, conceptualized as culture, and space as nature has come to connote in Western discourses the respective emblematics of masculinity and femininity. Interestingly, she overlooks how these ideas of time–culture–agency and space–nature–passivity are not merely genderized, they are over-determined racially. Eurocentric discourse is thoroughly anthropological, its utilization of an evolutionary conception of secular time constituted its self as 'modern' and its so-called other as primitive on the basis of a refusal of 'co-evalness', according to which dispersal in space reflected a sequence in time, difference was affirmed as distance (see Fabian 1983). This temporal nativization of the 'other', outside/inside the West is accompanied by a spatial nativization in which people are compressed into prefabricated landscapes, the village, the ghetto, the shanty town, and undergo a process of 'representational essentializing . . . in which one part or aspect of people's lives comes to epitomize them as a whole . . . in an anthropological taxonomy' (Clifford 1992: 100). Jardine's arguments nevertheless are useful in foregrounding these discourses in terms of: a 'reconceptualization of the act which has been the master narrative's own "non-knowledge", what has eluded them, what has engulfed them'. What I want to suggest is that 'This other than themselves' represents variable racialized articulations of times and spaces 'over which the narrative has lost control'. These are not only coded feminine, as Jardine argues, they are also coded ethnic–racial–other. Jardine uses the term 'gynesis' to describe the production in discourse of the 'woman effect' which signifies slippage between 'feminine', 'women', 'woman' (Jardine 1985: 24–8). Following her lead, I would like to rework the term 'Diasporics' in order to signify, if not capture, dynamic 'hi-life', 'jump-up', 'swing-beat', be-bop' time and space movements which are inscribed in the radical, iterability of the 'Black-identifying effects'. Its theorization is 'intrinsic to new and necessary modes of thinking, writing, speaking' (ibid.) and producing politics against the grain and in the grip of Western modernity's racially essentializing logics. In the words of Jardine 'The object produced by this process is neither a person or a thing, but a horizon, that toward which the process is heading' (Jardine 1985: 25). Diasporics describes a discursive journey inscribing destinations which are also the sites of departures, diasporic comings and goings, in which time and space must be thought in the moments before their separation. Derrida suggests the concepts of 'near and far' are both derivative of and prior to the opposition of time and space:

> The near and far are thought here, consequently, before the opposition of space and time, according to the opening of a spacing which belongs neither to time nor to space, and which dislocates, while producing it, any presence of the present.
>
> (Derrida 1982: 133)

Diasporics signifies the 'near and far' in Black experiences of racial antagonism and dislocation. It reveals the cultural contestation of the 'text milieu' (Hartman 1978), of the racialized, Western nation-state by invoking itself as 'extraterritorial' (Gates 1984). Through a Diasporics analysis and reading of the nation's times and spaces, a deconstruction of its political parochialism can be initiated through the cultural marking of unrepresentable limits and the politics of reading those limits.

BLACK JOURNEYING: A TEMPORARY CONCLUSION

In this concluding section I want to suggest how we might begin to contextualize the fragments of time-spatial analysis that I have tried to introduce in this chapter. In order to do this, I want to focus once again on my problematization of the postwar settlement period. It is possible to re-situate it as a 'chronotopic' (Bakhtin 1984) time-space fusion of continuity/discontinuity in regional and national 'ethnoscapes' (see below). Thus a reworking of the traditional race-relations narrative might produce the following as an alternative for analysis. Throughout the 1950s and 1960s the economic migration of Black people and others to Britain not only consolidated and elaborated regional racial antagonisms (hence re-affirming the identity of existing Black settlements) but also 'nationalized' the politics of race as a discourse about the racial configuration of the social: White society? Multicultural society? Movement into Britain was movement throughout Britain. The major cities, London, Birmingham, Manchester, Liverpool, Leicester, Bristol, experienced dislocations of 'local' identity as 'race' reinscribed the city cartography and, through a 'race-relations' sense of regional equivalence, a national, postwar, white racism emerged: televisual, legislative and class-unconscious.

The contours of Black settlement are, however, always more than residential, they are cultures of movement (see below). The articulation of community itself is a discursive investment in time-spatial constitutions within and beyond the nation-state. This not only inscribes pluralism as cultural difference in the social, but also articulates conditions in which struggles against racism in housing, education, policing and other social institutions are fought and thought as attempts to revalorize the terms upon which public justice has been settled. The affirmation of Black community is its politicization, even in a day-to-day sense. Categoric distinctions between public and private, personal and political, local and national, are blurred in the activities of diasporic identification. The heterographies of settlement, which narrativize the spatial and temporal institution of community, appropriate regional Britain both in iconographic and cartographic vernaculars: Handsworth (Birmingham), Brixton (London), Moss Side (Manchester), Chapeltown (Leeds), Liverpool 8 signify not merely the dispersed incidence of Black settlement but the traces of agency, assertion, harassment, local politics, recreational routes and cultural expressivity. The regionalism of Black identity as well as the geography of the Black presence inside/outside Britain expresses that identity, defines its politics and re-articulates Britain.

Having outlined an initial basis for rethinking the cultural politics of Black settlement, I want to discuss briefly two related perspectives which may assist in understanding further the racialized contexts of the 'chronotype', settlement. The first, 'population contexts', can be discussed in terms of 'ethnoscape' (Appadurai 1990, 1991; Hesse *et al.* 1992). Appadurai introduced the term 'global ethnoscape' to describe the 'landscape of persons who constitute the shifting world in which we live'. He characterizes the global ethnoscape as a moving world of tourists, immigrants, refugees, guestworkers, exiles. These populations form an essential feature of the world in political terms. Appadurai emphasizes the changing, social, territorial and cultural reproduction of group identity. Landscapes of group identity are penetrated by what lies outside them, they are no longer tightly territorialized. The world is increasingly traversed by groups who have to deal with the 'realities of having to move or the fantasies of wanting to move' (Appadurai 1991: 192). If understood as a conduit in the globalizing dissemination of distinctive forms of experience, this example of 'global culture' (see Featherstone 1990) is systematically racialized. For example, the 'fortress' of a more politically integrated, economically 'liberated' single Europe closes its doors even before the non-white or 'Third World traveller' has had time to be refused entry. As Balibar (1991: 6–7) has observed, while 'the present pattern of migrations does not inevitably "produce" racism – as a certain conservative discourse frequently maintains – it does give contemporary racism a focus'. The question of racism, Europe and population movements re-figures and unsettles the settling of Black (and Asian) communities in Britain quite profoundly:

> The fact that, in Europe as a whole, a large proportion of 'Blacks' or immigrants are not foreigners in the eyes of the law merely intensifies the contradictions, and intersects with the ever more pressing question of European identity. On the one hand, then, the emergence of a European racism, or model of racism, raises the issue of Europe's place in a world system, with its economic inequalities and population flows. On the other hand, it appears to be inextricably bound up with questions relating to collective rights, citizenship, nationality and the treatment of minorities, where the real political framework is not each particular country but Europe as such.
>
> (Balibar 1991: 6–7)

The racialized politico-cultural economy of Europe is a 'building block' of an 'imagined world' (Appadurai 1990: 298) in which Black and Asian communities in Britain become no more certain of moving freely in Europe and returning, than moving freely where we are: since when we move, 'race' moves. I have tried to develop the 'local' implications of this in my formulation of 'local ethnoscape' which follows Appadurai's lead (see Hesse *et al.* 1992: 167–9). This describes the landscape of persons who 'locally' connote the recurrent world on which lived experiences rely daily for definitions of 'place', where irrespective of long-term demographic transformations, transient permanence rather than permanent transience characterizes the moving population at any point in time. Only through the proportional recurrence of categories like 'race', gender, ethnicity, sexuality, class, age, (in their intra-fluidity and compositionality) can we understand where 'difference' is gathered from and how it is calculated as an investment in the imagination and territorialization of communities. In Britain, local 'racial antagonisms' work through 'transient congregations' (e.g. housing estate populations, school populations, street populations) in struggles to define the difference that 'race' makes to the space of the neighbourhood or the nation. What I have previously described as 'white territorialism' (Hesse *et al.* 1992: 173) motivates racial harassment as an affirmation of the exclusive dominance of white identities in the local ethnoscape.

The second contextual perspective to which I need to refer might be described as the 'context of movements'. I begin with Clifford (1992: 90–1, 104) whose discussion of 'travelling cultures' is particularly relevant. Clifford defines the 'chronotope of a culture' as a 'setting or scene organizing time and space in a representable form'. As a starting point for analysis, it is geared towards a 'notion of comparative knowledge produced though an itinerary, always marked by a way in; a history of locations and a location of histories' (ibid.: 105). This is yet another way of returning to the logics of Black settlement, our travelling cultures and our cultures of travel, by way of 'constructed and disputed historicities, sites of displacement, interference, interaction' (ibid.). Clifford's framework of analysis notes that travellers move under strong economic and political compulsion, and that privilege and oppression distinguishes different travellers, leading to different determinants of travel. In arguing for a comparative analysis of dwelling/travelling, Clifford wants to foreground 'discrepant cosmopolitanism', the outcome of unresolved historical dialogues between 'continuity and disruption', 'essence and position', 'homogeneity and difference' (ibid.). This highlights the unresolved discrepancies between slavery, colonialism, contemporary racism, and living in Britain. Although travelling cultural theory deconstructs the Eurocentric discourse of the 'West and the Rest' (Hall 1992), it is not without its difficulties for understanding Black experiences. bell hooks (1992) is quite sympathetic to Clifford's approach but reveals a number of 'blind spots' which are significant. The concept of travel, she suggests, is too closely tied to the imperialist adventures of European explorers, missionaries and dilettantes. New routes need to be uncovered rather than old ones rerouted if the representation of different journeys is to become possible:

> Theories of travel produced outside conventional borders might want the journey to become the rubric within which travel as a starting point for discourse is associated with different headings – rites of passage, immigration, enforced migration, relocation, enslavement, homelessness.
>
> (hooks 1992: 343)

hooks' alternative conception of 'journeying', theorized in diverse forms, is closely related to my understanding of Black heterographies of settlement in Britain: both are crucial to understanding any politics of locations. In these terms, journeys become movements in time, space and memory in reconstituting archaeologies of places of attachment or bewilderment, which are reinhabited in order to comprehend contemporary settings. As hooks points out, 'travel is not a word that can easily be evoked about the middle passage'. Once diverse journeys are grasped as cultures of movement, encounters with racism in the dwelling zone or challenges against the racialized accretions of the past are not so easily eclipsed by the 'hegemony of one experience of travel' (ibid.). Journeying is a powerful concept to employ in describing the political conditional logic of Black settlement in Britain. Journeying is as much sideways and in the standing still as it is 'backwards and forwards'. This describes the dilemmas, the uncertainties of Black settling, as racialized journeying in the city, in work, out of work, in school, in the street, on the estate, in the British nation, where, culturally and politically, the possibilities of inscription (space) and the possibilities of possibility (time) constitute a domain of struggle which presupposes all others.

ACKNOWLEDGEMENTS

I should like to thank Joanne Commings, Paul Gilroy, Michael Keith, Kwamla Hesse, Ernesto Laclau, Dhanwant K. Rai, and Bobby Sayyid – all of whom in their respective, different ways energized me to get this written.

REFERENCES

Aareen, R. (1989) *The Other Story*, London: Hayward Art Gallery.
Amin, S. (1988) *Eurocentrism*, London: Zed Books.
Anderson, B. (1983) *Imagined Communities*, London: Verso.
Appadurai, A. (1990) 'Disjuncture and difference in the global cultural economy', in M. Featherstone (ed.) *Global Culture*, London: Sage.
—— (1991) 'Global ethnoscapes', in R. G. Fox (ed.) *Recapturing Anthropology*, New Mexico: School of American Research Press.
Asante, M. (1987) *The Afrocentric Idea*, Philadelphia: Temple University.
—— (1988) *Afrocentricity*, Trenton, N.J.: Africa World Press Inc.
Bakhtin, M. (1991) *The Dialogic Imagination*, Austin: University of Texas Press.
Balibar, E. (1991) 'Racism and politics in Europe today', *New Left Review* 186.
Bhabha, H. (ed.) (1990) *Nation and Narration*, London: Routledge.
Burning Spear (1975) 'Slavery days' from LP 'Marcus Garvey', Island Records (London).
Clark, T. J. A. (1987): 'Time after time: temporality, temporalization', *The Oxford Literary Review* 9.
Clarke, V. A. (1991) 'Developing diaspora literacy and Marasa consciousness' in H. J. Spillers (ed.) *Comparative American Identities*, London: Routledge.
Clifford, J. (1992) 'Travelling cultures', in L. Grossberg, C. Nelson and P. Treichler, *Cultural Studies*, London: Routledge.
Dennis, F. (1988) *Behind the Frontlines*, London: Victor Gollancz Ltd.
Derrida, J. (1982) *Margins of Philosophy*, Brighton: Harvester Wheatsheaf.
Dreyfus, H. L. and Rainbow, P. (1986) *Michel Foucault*, Kent: Harvester Press.
Fabian, J. (1983) *Time and the Other: How Anthropology Makes its Object*, New York: Columbia University Press.
Fanon, F. (1963) *The Wretched of the Earth*, Harmondsworth: Penguin.
Featherstone, M. (ed.) (1990) *Global Culture*, London: Sage.
Foucault, M. (1972) *The Archaeology of Knowledge*, London: Tavistock Publications.
—— (1984) 'The order of discourse', in M. Shapiro (ed.) *Language and Politics*, Oxford: Blackwell.
Fryer, P. (1984) *Staying Power*, London: Pluto Press.
Gashe, R. (1986) *The Tain of the Mirror*, London: Harvard University Press.
Gates, H. L. (ed.) (1984) *Black Literature and Literary Theory*, London: Routledge.
Gilroy, P. (1987) *There Ain't No Black in the Union Jack*, London: Unwin Hyman.
—— (1991) 'It ain't where you're from, it's where you're at . . . the dialectics of diasporic identification', *Third Text* 13.
—— (forthcoming) *Promised Lands*, London: Verso.
Glissant, E. (1989) *Caribbean Discourse*, Charlottesville: University Press of Virginia.

Hall, S. (1978) 'Racism and reaction', in *Five Views of Multi-racial Britain*, London: CRE.
—— (1992) 'The West and the rest' in S. Hall and B. Gieben (eds) *Formations of Modernity*, London: Polity Press.
Harris, J. E. (ed.) (1982) *Global Dimensions of the African Diaspora*, Washington, DC: Howard University Press.
Hartman, G. (1978) 'A short history of practical criticism', *New Literary History* 10.
Hawkes, T. (1977) *Structuralism and Semiotics*, London: Routledge.
Hesse, B., Rai, D. K., Bennett, C. and McGilchrist, P. (1992) *Beneath the Surface – Racial Harassment*, London: Avebury.
hooks, b. (1992) 'Representing whiteness in the Black imagination', in L. Grossberg, C. Nelson and P. Treichler *Cultural Studies*, London: Routledge.
Jardine, A. (1985) *Gynesis: Configurations of Women and Modernity*, Ithaca, NY: Cornell University Press.
Laclau, E. (1990) *New Reflections on the Revolution of Our Time*, London: Verso.
Laclau, E. and Mouffe, C. (1985) *Hegemony and Socialist Strategy*, London: Verso.
Lawrence, E. (1982) 'Just plain common sense: the "roots" of racism' in Centre for Contemporary Studies *The Empire Strikes Back*, London: Hutchinson.
Little, K. (1948) *Negroes in Britain*, London: Routledge and Kegan Paul.
Lyotard, J.-F. (1984) *The Postmodern Condition*, Manchester: Manchester University Press.
Marable, M. (1985) *Black American Politics*, London: Verso.
Morrison, T. (1990) 'The site of memory' in R. Ferguson, M. Gever, T. T. Minh-Ha and C. West *Out There Marginalization and Contemporary Cultures*, Cambridge, Mass.: MIT Press.
Mudimbe, V. Y. (1988) *The Invention of Africa*, London: James Currey.
Pieterse, J. N. (1989) *Empire and Emancipation*, London: Pluto Press.
Ramdin, R. (1987) *The Making of the Black Working Class in Britain*, Aldershot: Wildwood House.
Robinson, C. (1983) *Black Marxism*, London: Zed Press.
Said, E. (1978) *Orientalism*, London: Penguin.
Scott-Heron, G. (1977) '95 South (All of the places we've been)', from the LP 'Bridges', Arista (London).
Small, S. (1991) 'Racialised relations in Liverpool: a contemporary anomaly', *New Community*, July, 17(4): 511–37.
Snead, J. A. (1984) 'Repetition as a figure of black culture', in H. L. Gates (ed.) *Black Literature and Literary Theory*, London: Routledge.
Soja, E. (1989) *Postmodern Geographies*, London: Verso.
Solomos, J. (1989) *Race and Racism in Contemporary Britain*, London: Macmillan.
Spivak, G. (1992) 'Asked to talk about myself', *Third Text* 19.
Staten, H. (1984) *Wittgenstein and Derrida*, Oxford: Basil Blackwell.
Steel Pulse (1982) 'Rally Round' from LP 'True Democracy', Wiseman Doctrine Records (London).
Thompson, V. B. (1987) *The Making of the African Diaspora in the Americas 1441–1900*, New York: Langasan.
Todorov, T. (1982) *The Conquest of America*, New York: Harper & Row.
Vattimo, G. (1988) *The End of Modernity*, Cambridge: Polity Press.
West, C. (1988) 'Marxist theory and the specificity of Afro-American oppression', in C. Nelson and Grossberg (eds) *Marxism and the Interpretation of Culture*, London: Macmillan.
Wheeler, C. (1990) 'UK Blak', from LP of same name, RCA Records (UK).

Wolfe, E. R. (1982) *Europe and the People without History*, London: University of California Press.
Wood, D. (1989) *Deconstruction of Time*, Atlantic Highlands, NJ: Humanities Press International.
Wright, P. (1985) *On Living in an Old Country*, London: Verso.
Zhana (1988) *Sojourn*, London: Methuen.

10

THE SPACES THAT DIFFERENCE MAKES

Some notes on the geographical margins of the new cultural politics

Edward Soja and Barbara Hooper

> Postmodern culture with its decentred subject can be the space where ties are severed or it can provide the occasion for new and varied forms of bonding. To some extent, ruptures, surfaces, contextuality, and a host of other happenings create gaps that make space for oppositional practices which no longer require intellectuals to be confined to narrow separate spheres with no meaningful connection to the world of the everyday . . . a space is there for critical exchange . . . [and] this may very well be 'the' central future location of resistance struggle, a meeting place where new and radical happenings can occur.
>
> (bell hooks 1990: 31)

> These gestures are not new in the history of criticism . . . yet what makes them novel – along with the cultural politics they produce – is how and what constitutes difference, the weight and gravity it is given in representation and the way in which highlighting issues like exterminism, empire, class, race, gender, sexual orientation, age, nation, and region at this historical moment acknowledges some discontinuity and disruption from previous forms of cultural critique. To put it bluntly, the new cultural politics of difference consists of creative responses to the precise circumstances of our present moment . . . in order to empower and enable social action.
>
> (Cornel West 1990: 20)

These introductory quotations, from two leading African-American cultural critics and philosophers, raise challenging questions about the specificity and the generality of current debates on the new cultural politics of difference. Both bell hooks and Cornel West are primarily concerned with reconceptualizing radical African-American subjectivity in a way that retains and enhances the emancipatory power of blackness, but is at the same time innovatively open to the formation of multiple communities of resistance, polyvocal political communities capable of linking together many radical subjectivities and creating new 'meeting places' and 'spaces' for diverse oppositional practices. Their explicitly postmodern viewpoint concentrates on the role of the African-American intellectual (as cultural critic and philosopher) but searches for wider spheres of participation in the 'world of the everyday' and in the enablement of radical social action everywhere in the world, from the local to the global.

Can these outlooks and strategies, so deeply rooted in the particular circumstances of the African-American experience, be taken more generally as a programme and strategy for a radical postmodern cultural politics of difference? Might such adaptation by others who do not directly share this experience be seen as a form of co-optation, opportunism, or misplaced idealism? Will these debates themselves provoke a response that they are merely intellectual exercises, divorced and abstracted from 'real' radical politics? We begin our discussion with a 'yes' to all three questions, the first as a positive assertion of a radical postmodernism that we share with hooks and West as, in hooks' words, a 'fertile ground for the construction of empathy';[1] and the second and third as acknowledged dangers and tensions that accompany the new possibilities opened up in the discourse and practice of a radical postmodern cultural politics of difference.[2]

We will explore, in admittedly broad-brush fashion, the new cultural politics of difference first in a discussion of the differences that postmodernity makes: why what is 'new' is so closely attached to 'postmodern culture' and the 'precise circumstances of our present moment'. We move then to the spaces that difference makes: why the critical discourse of radical postmodernism is so often framed in spatial terms and why this discursive spatiality should not be interpreted as merely metaphorical, a newly fashionable trope that is disconnected from any material geographical reality. Instead, we suggest that this spatialized discourse on simultaneously real and imagined geographies is an important part of a provocative and distinctively postmodern reconceptualization of spatiality that connects the social production of space to the cultural politics of difference in new and imaginative ways. We conclude with some critical comments on the implications of the new postmodernized and spatialized cultural politics of difference for the academic discipline of geography.

ON THE DIFFERENCES THAT POSTMODERNITY MAKES

The cultural politics of difference, whether old or new, arise primarily from the workings of power – in society and on space in both their material and imagined forms. Hegemonic power does not simply manipulate naïvely given differences between individuals and social groups, it actively *produces and reproduces difference* as a key strategy to create and maintain modes of social and spatial division that are advantageous to its continued empowerment. At the same time, those subjected, dominated, or exploited by the workings of hegemonic power and mobilized to resist by their putative positioning, their assigned 'otherness', struggle against differentiation and division.[3] This socio-spatial differentiation, division and struggle is, in turn, cumulatively concretized and conceptualized historically and geographically as *uneven development*, a term which we use to describe the composite and dynamic spatio-temporal patterning of socially constructed differences at many different scales, from the local to the global.

Such differences as are ascribed to gender, sexual practice, race, class, region, nation, etc. are thus primarily 'brute fashionings' (Cocks 1989: 20) which are neither transhistorical nor 'natural' (in the sense of being naïvely or existentially given). This brute fashioning, as the social production (and strategic reproduction) of difference, is the catalyst for both hegemonic (conservative, order-maintaining) and counter-hegemonic (resistant, order-transforming) cultural and identity politics. That the historical process of uneven development is also intrinsically spatial as both medium and outcome gives to the (real and imagined) geography of cultural politics a particular significance.[4]

Counter-hegemonic cultural politics has usually taken two broad forms, at least within Western capitalist societies. The first, rooted in the post-Enlightenment development of liberal humanism, has traditionally based its opposition on the assertion of universal principles of equality and democracy, seeking to reduce to a minimum the negative effects of difference whatever their origins. A second form of counter-hegemonic politics, not always completely separable from the first, arises from more radical contestation over the many axes along which socially constructed power differentials are polarized. Without excluding the liberal alternative entirely, what we define here as *modernist identity politics* refers primarily to this tradition of more radical subjectivity and resistance as it has developed since the mid-nineteenth century around such categories of cultural consciousness as class, race, ethnicity, nationality, colonial status, sexuality and gender.

Even when rejecting Marxian categories and explicitly revolutionary ideology, modernist identity politics and the various social and cultural movements associated with it have characteristically tended to develop along the lines charted out more than a century ago with respect to the formation of anti-capitalist class-consciousness and the revolutionary struggle against exploitation. While varying greatly in their specificities, these movements have generally followed analogous trajectories based on a similar praxis of refusal and resistance that parallels the bipolar logic of class struggle: capital vs. labour, bourgeoisie vs. proletariat: a struggle defined around a deep structural dichotomy that 'orders' differential power into two primary social categories, one dominant the other subordinate.[5]

Each separate sphere of modernist identity politics has typically mobilized its version of radical subjectivity around a fundamentally epistemological critique of the binary ordering of difference that is particular to it: capital/labour, self/other, subject/object, colonizer/colonized, white/black, man/woman, majority/minority, heterosexual/homosexual. The critique is aimed at 'denaturalizing' the origins of the binary ordering to reveal its social and spatial construction of difference as a means of producing and reproducing systematic patterns of domination, exploitation and subjection. As socially constructed, context-specific 'technologies' of power (to use Foucault's term), these binary structures become subject to social and cultural transformation via a politics of identity that builds upon the empowerment of the 'subaltern' against the 'hegemon' (to use the most general terms covering the various oppositional forms of the cultural politics of difference).[6]

Modernist identity politics characteristically projects its particular radical subjectivity, defined within its own oppressive binary structure, as overarchingly (and often universally) significant. Whether or not this totalization and essentialism is actually believed, its powerful mobilizing effect is used strategically in attempts to consolidate and intensify counter-hegemonic consciousness 'for itself' and on its identified 'home ground'. In both theory and practice, therefore, a significant degree of closure and exclusiveness is embedded within the strategies and tactics of modernist identity politics. Even when one form avows its openness to alliance with others, it is usually open only on the former's terms and under its primary strategic guidance. The result has been the production of parallel, analogous, but rarely intersecting channels of radical political consciousness, each designed and primed to change their own discrete binary world of difference.[7]

Under these ordered conditions, *fragmentation* (in the very real form of complex multiple subjectivities, with and without overlapping) becomes an endemic problem in modernist identity politics, especially for those social movements that theorize either a universalist encompassing of other radical subjectivities (e.g. substituting woman for women or the transcendental unity of an international working class for multiracial, multi-ethnic, and otherwise diverse men and women workers) or, alternatively, that recognize differences but none the less theorize and strategize from the assumption of the primacy and privileging of one or another set of agents in the process of radical social transformation. In the extreme case, as with most orthodox forms of modern Marxism and some forms of radical feminism and black nationalism, these essentialist tendencies abrogate any cross-cutting alliances of political significance by attributing 'false consciousness' or subordinate identity to all radical subjectivities other than that emanating from the 'primary' bipolarity.[8]

When the primacy of one binary is viewed as competing with the privileging of another, the prospects for flexible and co-operative alliance and empathy are likely to be dim. While there have been fruitful dialogues between various radical movements (between Marxism and feminism, for example, and between both and those struggling against racism and colonial oppression), the deeply engrained essentialisms of modernist identity politics have tended to create a competitive exclusivity that resists, even rejects, seeing a 'real' world populated by multiple subjects with many (often changeable) identities located in varying (and also changeable) subject positions. Hence, modernist identity politics, in its fear and rejection of a fragmented reality, has often tended to create and intensify political divisiveness rather than working toward a multiple, pluralized, and yet still radical conceptualization of agency and identity.

Modernist identity politics has always had to face reactionary hegemonic resistance and reformist liberal diversion. Over the past two decades, however, new theoretical and philosophical critiques (especially of taken-for-granted epistemologies and, in particular, the politically divisive tendencies toward master-narrative essentialism and binary totalization) and actual political events (ranging from the global restructurings of capital, labour, and ideology to the apparent defeat and retreat of governments inspired by modernist Marxism-Leninism and its dominant form of identity politics) have ushered in an extraordinary period of deep questioning that strikes more disruptively than ever before at the very foundations of modernist political practices – and simultaneously opens up new possibilities for radical resistance to all forms of hegemonic subordination.

There is no doubt that some who hasten to proclaim the death of modernism and announce the emergence of a new postmodern era from its ashes are motivated by the same old impulses of hegemonic re-empowerment and liberal diversion. But there is also emerging – in the postmodern blackness and post-colonialist critiques of bell hooks, Cornel West, Gayatri Spivak, Arjun Appadurai, Trinh T. Minh-Ha, Edward Said, Chandra Mohanty, Homi Bhabha; in the postmodern feminisms of Iris Marion Young, Jane Flax, Judith Butler, Diana Fuss, Meaghan Morris, Rosalyn Deutsche, Donna Haraway; and in the anti-essentialist critiques of various postmodern Marxist scholars such as Ernesto Laclau and Chantal Mouffe – a polyvocal postmodernism that maintains a commitment to radical social change while continuing to draw (selectively, but sympathetically) from the most powerful critical foundations of modernist identity politics. The intent behind this radical postmodernism of resistance is to deconstruct (not to destroy) the ebbing tide of modernist radical politics, to renew its strengths and avoid its weaknesses, and to reconstitute an explicitly postmodernist radical politics, a new cultural politics of difference and identity that moves toward empowering a multiplicity of resistances rather than searches for that one 'great refusal', the singular transformation to precede and guide all others.

The *disordering* of difference from its persistent binary structuring and the reconstitution of difference as the basis for a new cultural politics of multiplicity and strategic alliance among all who are peripheralized, marginalized and subordinated by the social construction of difference (especially in its binary forms) are key processes in the development of radical postmodernism.

Whether this revisioning of radical subjectivity requires a major transformation or merely a significant reform of modernist identity politics is still being contested. But it is clear that politics as usual can no longer be practised as it was in the past, at least among those who take seriously the conditioning effects of postmodernity.[9]

In the wake of this continuing debate, a new breed of radical anti-postmodernists has emerged in force to 'spin-doctor' the critical discourse toward reformist solutions and away from any deep deconstruction and reconstitution of modernist traditions. Given the intellectual elitism and neoconservative politics that have dominated so much of the postmodernist discourse, there is ample ammunition for the radical anti-postmodernist project. After all, the deconstructive challenge raised by postmodernism often sounds suspiciously like the same old hegemonic strategies of opposition, co-optation, and diversion; and the affirmedly prefixed 'post' seems too literally to signal the irrevocable end of all progressive modernist projects rather than their potentially advantageous reconstitution. Moreover, to most modernist critics, the multiplicity of resistances continue to be seen as inevitably leading to a politically debilitating fragmentation and the abandonment of long-established forms of struggle. Under these presumed circumstances, the promise of eventual emancipatory reconstitution rings hollow, if not cruelly deceptive, especially at a time when nearly all radical modern movements are either in crisis or massive retreat.

The tendency to homogenize postmodernism and to totalize (i.e., ascribe to all postmodernisms) certain negative and oppressive political practices associated with postmodernity, makes the construction of a radical postmodern alternative even more difficult.[10] Theoretically suggestive aphorisms arising from the postmodern and related post-structuralist critiques ('there is no reality outside the text', 'the death of the subject', 'anything goes') are now routinely set up as straw-objects to prove that all postmodern politics are abstract, unrelated to everyday life, inherently reactionary – and hence, immanently nihilistic with respect to real radical politics. A forbidding wall of categorical totems and taboos has thus been raised to hide the very possibility of radical postmodernism, making it all but invisible to outsiders, especially on the left.

In our view, this forbidding wall has materialized around the same modernist conditioning and rigid either/or logic that has become so central a target for contemporary cultural criticism and the new politics of difference: the infatuation with clean and orderly binary oppositions; the intolerance of ambiguity, disordering, multiplicity, fragmentation; the urge to unity enforced by epistemological closure and essentialism. The arguments we have outlined as a critique of modernist identity politics are thus a critique of the flourishing new anti-postmodernism. It is a critique that calls for a new way of looking at and making practical political sense of the precise circumstances of the contemporary moment, drawing insight from the realization that postmodernity has made more significant differences to our real and imagined political worlds than it has reinforced continuities with the past.

To turn what has thus far been a primarily deconstructive critique of modernism into a more reconstructive programme for a radical postmodernist cultural politics requires developing a more complex understanding of the key themes of disordering difference and empowering multiplicity. To work toward this understanding, our discussion takes a more explicitly spatial turn, into a subregion of the contemporary literature on the new cultural politics of difference that has begun to reconceptualize its discourse and critique around the spaces that difference makes. We suggest that it is here, in the creative spatialization arising from the broad field of cultural studies, that radical postmodernism is being most effectively conceptualized and made practically political.

THE SPATIAL TURN IN THE NEW CULTURAL POLITICS OF DIFFERENCE

> As a radical standpoint, perspective, position, 'the politics of location' necessarily calls those of use who would participate in the formation of counter-hegemonic cultural practice to identify the spaces where we begin the process of re-vision. . . . For me this space of radical openness is a margin – a profound edge. Locating oneself there is difficult yet necessary. It is not a 'safe' place. One is always at risk. One needs a community of resistance.
>
> (hooks 1990: 145–9)

In *Yearning: Race, Gender, and Cultural Politics*, and particularly in the chapters 'Postmodern blackness' and 'Choosing the margin as a space of radical openness', bell hooks (1990) attempts to move beyond modernist binary oppositions of race, gender and class into the multiplicity of *other* spaces that difference makes; into a re-visioned spatiality that creates from difference new sites for struggle and for the construction of interconnected communities of resistance. In so doing, she opens up in these real and imagined 'other spaces' the possibilities for a new cultural politics of difference and identity that is both radically postmodern and strategically spatialized from the start. This creative (re)spatialization is more than an appealing metaphor or abstraction. It is a vital discursive turn that both grounds the new cultural politics and facilitates its conceptual re-visioning around the empowerment of multiplicity, the construction of combinatorial rather than competitively fragmented and separated communities of resistance.

Early work on the spatialization of cultural politics can be found in the writings of Siegfried Kracauer and Walter Benjamin, Franz Fanon and Simone de Beauvoir, Michel Foucault and Henri Lefebvre. In addition, there has developed over the past decade a substantial body of literature on gender, race, class, erotic preference and anti-colonialism that has been specifically concerned with disordering modernist binaries and promoting a new and radical cultural politics. In bell hooks' writings, there has been a particularly lucid, expressive, and accessible convergence of these discourses, making her *Yearnings* an especially useful place to begin the present discussion.

hooks finds her place, positions herself (first of all as an African-American woman), by the simultaneously political and geographical act of *choosing marginality*. This positioning of identity is detached from the 'narrow cultural nationalism masking continued fascination with the power of the white [and/or male] hegemonic order'. Such an identification, hooks suggests, would not be choosing marginality but accepting its imposition by the more powerful, binary Other, a submission to the dominant, order-producing, and unremittingly modernist ideology and epistemology of difference. Instead, such an assertion of recentred identity 'is evoked as a stage in a process wherein one constructs radical black subjectivity' (or what she has more recently called 'wildness' (hooks 1992)).[11] By extension and adjustment, choosing marginality can also become a critical turning point in the construction of other forms of counter-hegemonic or subaltern identity and more embracing communities of resistance.

As an initial stage, categorical identity as subaltern is crucial. But hooks' construction of radical black subjectivity pushes the process of identity formation beyond exclusionary struggles against white racism on to a new terrain, a 'space of radical openness' where the key question of 'who we can be and still be black' can be politically re-imagined. 'Assimilation, imitation, or assuming the role of rebellious exotic are not the only options and never have been', hooks notes in rejecting the conventional choices that liberal modernist discourse has frequently imposed upon the activist black subject. Instead, she chooses a space that is simultaneously central and marginal (and purely neither at the same time), a difficult and risky place on the edge, in-between, filled with contradictions and ambiguities, with perils but also with new possibilities.

'Fundamental to this process of *decentering the oppressive other* and claiming our right to subjectivity', she writes, 'is the insistence that we must determine how we will be and not rely on colonizing responses to determine our legitimacy.' In a similar avowal, Pratibha Parmar argues that creating identities as black women is not done ' "in relation to", "in opposition to", or "as a corrective to" . . . *but in and for ourselves*. Such a narrative thwarts that binary hierarchy of centre and margin: the margin refuses its place as "Other".' Significantly, hooks cites Parmar in 'Choosing the margin' to confirm their shared presupposition of the prolitical significance of spatiality. After stating that spaces 'can be real and imagined . . . can tell stories and unfold histories . . . can be interrupted, appropriated, and transformed', hooks adds: 'As Pratibha Parma[r] notes, "The appropriation and use of space are political acts" ' (see Parmar 1990: 101).

This alternative process of choosing marginality reconceptualizes the problematic of subjection (see note 3) by deconstructing both margin and centre, while reconstituting in the restructured (recentred) margins new spaces of opportunity, the new spaces that difference makes. For hooks, and by extension and invitation, all others involved in this spatial disordering of difference, there is a 'definite distinction between the marginality which is imposed by oppressive structure and that marginality one chooses as site of resistance, as location of radical openness and possibility'. She adds, more extensively:

> It was this marginality that I was naming as a central location for the production of a counter-hegemonic discourse that is not just found in words but in habits of being and the way one lives. As such, I was not speaking of a marginality one wishes to lose – to give up or surrender as part of moving into the centre, but rather as a site one stays in, clings to even, because it nourishes one's capacity to resist. It offers the possibility of radical perspectives from which to see and create, to imagine alternatives, new worlds. . . .
>
> Understanding marginality as position and place of resistance is crucial for oppressed, exploited, colonized people. If we only view the margin as a sign, marking the condition of our pain and deprivation, then certain hopelessness and despair, a deep nihilism penetrates in a destructive way. . . . I want to say that these margins have been both sites of repression and sites of resistance. . . .
>
> This is an intervention. A message from that space in the margin that is a site of creativity and power, that inclusive space where we recover ourselves, where we move in solidarity to erase the category colonizer/colonized. Marginality is the space of resistance. Enter that space. Let us meet there.
>
> (hooks 1990: 149–52)

In choosing marginality as a space of radical openness, hooks contributes significantly to a powerful revisioning not only of the cultural politics of difference but also of our conceptualization of human geographies, of what we mean by the politics of location and geohistorically uneven development, of how we creatively combine spatial metaphor and spatial materiality in an assertively spatial *praxis*. By recontextualizing spatiality, she engages in a cognitive remapping of our many real and imagined worlds – from the most local confines of the body, the spatiality closest in, to the nested geographical worlds that are repeated again and again in an expanding sequence of scales reaching from the 'little tactics of the habitat' to the 'great strategies' of global geopolitics.[12] For hooks, the political project is to occupy these (real and imagined) spaces on the margins, to reclaim them as locations of radical openness and possibility, and to make within them a place where one's radical subjectivity can be *seen* and practised in association with other radical subjectivities. It is thus a spatiality of inclusion rather than exclusion, a spatiality where radical subjectivities can multiply, connect and combine in polycentric communities of identity and resistance: the spatiality searched for but never effectively discovered in modernist identity politics.

The particular spaces that hooks explores can be gleaned from the Contents page of *Yearning:* 'Homeplace: a site of resistance'; 'The Chitlin circuit: on black community'; 'An aesthetic of blackness: strange and oppositional' (where she remembers her grandmother's house and the 'frightening' questions her Baba asked her: 'Do you believe that space can give life, can take it away, that space has power?); 'Aesthetic inheritances: history worked by hand' (a sensitive geography of the work and world of black women quilt-makers); 'Liberation scenes: speak this yearning' (an appreciation of Lorraine Hansberry's *A Raisin in the Sun* as a play 'about housing', about how 'racial segregation in a capitalist society' discriminates against blacks 'seeking a place to live'); 'Postmodern blackness' (where she finds 'new avenues to transmit the messages of black liberation struggle'); and, of course, 'Choosing the margin as a place of radical openness', the place where she makes her new spatiality most explicit.

In these spaces that difference makes for bell hooks, the distinctions between real and imaginary spaces and places, between spatial metaphors and materialized geographies, dissolve emphatically into what might be described as a 'thirdspace' of political choice, containing more than simple combinations of the original dualities. Although the phrase is not used by hooks, she enters directly into the remapping, the reconstitution, of *postmodern geographies*, both in the form of identifying revealing geographical 'texts' and 'scenes' that open our eyes to the politics of difference; and in developing new ways to read these geographies critically, to deconstruct and reconstitute them as spaces of radical opportunity, spaces that are simultaneously real, imagined and more.

Hooks has certainly not been alone in taking a critical spatial turn. Some of the predominantly white male and often Marxist antecedents of this critical reconceptualization of postmodern geographies – James Joyce, Georg Simmel, Walter Benjamin, Henri Lefebvre, Michel Foucault, John Berger, Fredric Jameson, David Harvey – have been noted here and elsewhere. Taking the lead more recently has been a group of predominantly feminist and often non-white cultural critics, who have revitalized the spatial turn and taken it into new grounds.

Remapping the city as a space of radical openness, a place where, like hooks' margin, ties are severed and subjection abounds but also, at the same time, a location for recovery and resistance, a meeting place where new and radical happenings can occur, has been a longstanding feminist project dating back at least to the fourteenth century.[13] The feminist urban critique, led by architects, planners and geographers, broadened its base in the late 1970s around revealing analyses of the gendering of space (especially in the forms of the urban built environment) and the reproduction of this male-dominated gendering through the contextualizing effects of patriarchy.[14]

For the most part, this expanded feminist critique and spatial remapping remained modernist, in the sense of channelling its critical power and emancipatory objectives around the gendered binary men/women. Urban spatiality thus came to be seen as oppressively gendered in much the same way that the city was shown to be structured by the exploitative class relations of capitalism and the discriminatory geographical effects of racism, the two other major channels of radical modernist urban critique developing over the same period. More recently, however, primarily through the development of an openly postmodern feminism, new directions are being taken that bring the feminist urban critique and remapping from its revealing recognition that space makes a difference into a creative exploration of the multiplicity of spaces that difference makes. What is new about this development, that is, how the postmodern feminist critique of urban spatiality can be distinguished from its modernist forms, brings us back to where we started, to 'how and what constitutes difference, the weight and gravity it is given in representation', as Cornel West argued; into a more encompassing politics of difference and identity that opens new spaces for critical exchange and creative responses 'to the precise circumstances of our present moment'; and into the strategic acts of disordering difference and choosing marginality.

In this growing literature,[15] many new ways of looking at urban spatiality are explored. The body as the most intimate of spatialities is rediscovered and given a central place in the construction of real and imagined geographies of the city, while through this embodiment the city becomes charged with multiple sexualities. The alternative spaces of the visual and aesthetic imagination – in films, photography, advertising, fashion, museum exhibitions, murals, poems, novels, but also in shopping malls and beaches, factories and streets, motels and theme parks – are imaginatively evoked as ways of seeing the city; and in the works of such cultural critics as Rosalyn Deutsche and Meaghan Morris, the construction of these spaces (as *oeuvres*, works of art, as well as *produits*, manufactured products) is connected directly to the material dynamics of geographically uneven development and the political economy of contemporary capitalism.[16] Iris Marion Young and Nancy Fraser, among others, have engaged effectively in deconstructing and reconstituting the ideology of urban community and the old modernist binary of public vs. private space in an explicitly radical postmodern and feminist politics of difference, while Donna Haraway has added nature and high technology to this ideological deconstruction and reconstitution.[17]

Of particular importance here and now, in these 'other' spaces and different geographies being opened up by the postmodern cultural critics, is the insight provided on how fragmentation, ruptures and discontinuities can be politically transformed from liability and weakness to opportunity and strength, a project which helps define the boundary between adaptive modernism and creative postmodernism. In the new postmodern cultural and geographical politics of difference, we position ourselves first by subjectively choosing *for ourselves* our primary 'marginal' identities as feminist, black, radical socialist, anti-colonialist, gay and lesbian activist. But we do not remain rigidly confined by this 'territorial' choice, as was usually the case in modernist identity politics. We seek instead to find more flexible ways of being other than we are while still being ourselves, of becoming open to combinations of radical subjectivities, to a multiplicity of communities of resistance, to what Trinh T. Minh-ha has called 'the anarchy of difference' (Trinh 1991: 120; see also Trinh 1989).

Trinh develops a strategy of *displacement* (as opposed to a strategy of reversal), which adds significantly to hooks' reconceptualization of marginality. 'Without a certain work of displacement', she writes, 'the margins can easily recomfort the center in goodwill and liberalism.' The margins are 'our fighting grounds' but also 'their site for pilgrimage . . . while we turn around and claim them as our exclusive territory, they happily approve, for the divisions between margins and center should be preserved, and as clearly demarcated as possible, if the two positions are to remain intact in their power relations'. By actively displacing and disordering difference, by insisting that there are 'no master territories', one struggles to prevent 'this classifying world' from exerting its ordered, binary, categorical power.

Diana Fuss adds further insight to the caution needed and dangers involved in choosing marginality, and helps to defend this strategic choice against those who feel uneasy when persons of substantial power and status (bell hooks? Ed Soja?) subjectively position themselves in the margins.

> Does inhabiting the inside always guarantee cooptation? . . . And does inhabiting the outside always and everywhere guarantee radicality? The problem, of course, with the inside/outside rhetoric, if it remains undeconstructed, is that such polemics disguise the fact that most of us are both inside and outside at the same time. Any displaced nostalgia for or romanticization of the outside as a privileged site of radicality immediately gives us away, for in order to realize the outside we must already be, to some degree, comfortably on the inside. We really only have the leisure to idealize the subversive potential of the power of the marginal when our place of enunciation is quite central.
>
> (Fuss 1991: 5)

Gayatri Chakravorty Spivak faces the challenge of choosing marginality and the locational politics of difference through a similar deconstruction, displacement and repositioning. But more so than hooks and the urban feminists, Spivak and other key figures in what has come to be called post-colonial or subaltern studies, move beyond deconstruction and repositioning to begin another remapping, one of more explicitly global proportions. For Spivak, this remapping is a move 'beyond a homogeneous internationalism, to the persistent recognition of heterogeneity', to a new 'worlding of the world' (Spivak 1988a: 20; see also Spivak 1988b). She encapsules the difficulty of this task in the slow and careful labour of 'un-learning our privilege as our loss', of 'behaving as if you are part of the margin' but doing so by relinquishing the privileges that attach to choosing marginality as a space of radical resistance from the centre (Spivak 1990: 9, 30).

Thinking synchronically, in the precise (spatial) circumstances of the present (postmodern) moment, Spivak positions herself as a *bricoleur*, a preserver of discontinuities, an interruptive critic of the categorical logic of colonizer–colonized, elite–subaltern, global–local, centre–periphery, First World–Third World.

> If I have learned anything it is that one must not go in the direction of a Unification Church, which is too deeply marked by the moment of colonialist influence, creating global solutions that are coherent. On the other hand, it seems to me that one must also avoid as much as possible, in the interests of practical effectiveness, a sort of continuist definition of differences, so that all you get is hostility.
>
> (Spivak 1990: 15)

Spivak's emphasis on discontinuities pushes us away from the simple celebration of pluralism and multiculturalism that has attracted so many contemporary liberal scholars; but at the same time it sets clear limits upon her desired preservation of the discourses of feminism and Marxism. What must be preserved, Spivak argues, is not the continuist definition of differences embedded in the coherent modernist discourses, the unified parochialism that responds to 'others' only with hostility; but rather a 'radical acceptance of vulnerability' (ibid.: 18) immersed in a deconstructive *pratique sauvage*, a 'wild practice' (ibid.: 54) a 'politics of the open end' (ibid.: 104) in which choosing marginality becomes an invitingly anarchic (without a chief, head, or authoritative centre) act of inclusion.

Spivak's remapping (reworlding) disrupts the coherent spatiality of territorial imperialism, Eurocentrism, and spatial science – as well as the spatiality of modernist feminism and Marxism. And she does so with a strategic twist that brings 'hegemonic historiography to crisis', a critique of Western historicism that becomes a vital turn in her reworlding project and in the larger spatialization of the new cultural politics of difference. Edward Said takes this critique of historicism still further in his deconstruction of the binary Orientalism–Occidentalism (see Said 1979; 1985) and critique of the 'imaginary geographies' historically constructed from the concrete, experienced, material geographies of imperialism.

> So far as Orientalism in particular and European knowledge of other societies in general have been concerned, historicism meant that one human history uniting humanity either culminated in or was observed from the vantage point of Europe, or the West. . . . What . . . has never

> taken place is an epistemological critique at the most fundamental level of the connection between the development of a historicism which has expanded and developed enough to include antithetical attitudes such as . . . critiques of imperialism on the one hand [including the world histories of Fernand Braudel, Immanual Wallerstein, Perry Anderson, Eric Wolf], and on the other, the actual practice of imperialism by which accumulation of territories and population, the control of economies, and the incorporation and homogenization of histories are maintained.
>
> [What we require] is a fragmenting, dissociating, dislocating and decentering of the experiential terrain covered by universalizing historicism . . . [that goes] beyond the polarities and binary oppositions of marxist-historicist thought . . . in order to create a new type of analysis of plural, as opposed to single, objects . . . We cannot proceed therefore unless we dissipate and re-dispose the material of historicism into radically different objects and pursuits of knowledge, and we cannot do that until we are aware clearly that no new projects of knowledge can be constructed unless they fight to remain free of the dominance and professionalized particularism that comes with historicist systems and reductive, pragmatic, and functionalist theories.
>
> (Said 1985: 22–3)

The head-on critique of historicism adds the finishing touch to our discussion of the spatial turn taken in the new cultural politics of difference. While neither Spivak nor Said explicitly connect their critique of historicism to this contemporary spatialization, they go much further than most of the other cultural critics and philosophers we have mentioned in making the deconstruction of historicism a prerequisite for all new projects, for radically different objects and pursuits of knowledge, and for this reassertive spatial turn.

FROM MODERN GEOGRAPHY TO POSTMODERN GEOGRAPHIES

> [A] growing skepticism concerning older explanatory models based in history has led to a renewed interest in the relatively neglected, 'under-theorized' dimension of space. . . . It has become less and less common in social and cultural theory for space to be represented as neutral, continuous, transparent or for critics to oppose 'dead . . . fixed . . . undialectical . . . immobile' space against the 'richness, fecundity, life, dialectics' of time, conceived as the privileged medium for the transmission of the 'messages' of history. Instead spatial relations are seen to be no less complex and contradictory than historical processes, and space itself refigured as inhabited and heterogeneous, as a moving cluster of points of intersection for manifold axes of power which cannot be reduced to a unified plane or organized into a single narrative.
>
> (Hebdige 1990)[20]

> The agendas of the new cultural politics deploy a richly spatialised vocabulary that should render a dialogue with geography in understanding the absences and presences of such representations of space mutually indispensable. However . . . ma(i)nstream geographical thought has been quarantined from their influence. . . . Geography as an academic discipline is in danger of becoming an anachronism, without a language to articulate the new space of resistance, the new politics of identity.[21]

Are the postmodern geographies being presented by the new cultural critics and philosophers significantly different from the Marxist, feminist, anti-racist and other radical human geographies currently being produced within the modern discipline that makes 'Geography' its nominal focus of intellectual identity? Everyone involved might now agree that 'geography matters' and that 'space makes a difference' in theory, culture and politics; but does this agreement mask conflicting perspectives on the 'spaces that difference makes' and hence on the new cultural politics?

These are complex questions that cannot be fully answered here. In any case, to create too great a gap between geographers and non-geographers is politically unwise and would obscure the exciting cross-fertilization that is currently taking place among those convinced of the political, theoretical and discursive significance of the spatiality of human life. Instead of engaging in a divisive polemic, therefore, we will conclude by addressing two arguments that directly reflect what many radical (and other) geographers might find useful from the spatial explorations outlined in the previous pages, leaving to others the equally interesting question of what the new spatializers can learn from geographers.

Behind the first argument is the suggestion that hooks' real and imagined spaces of critical dialogue, disordered marginality, and communities of resistance, as well as many of the other explorations of the spaces that difference makes highlighted in this essay, and including the influential spatializations of Michel Foucault and Henri Lefebvre, represent efforts to reach beyond the two predominant modes of spatial thinking that have defined and confined the geographical imagination for centuries. Reflecting perhaps the biggest of all philosophical dichotomies, that between subject and object, these two modes can be broadly described as subjectivist and objectivist.

The objectivist or materialist mode of spatial thinking and analysis has dominated modern geography throughout the twentieth century in a sequence that has shifted emphasis several times – from the simple contingencies of environmental determinism and possibilism, to the theoretically innocent description of the areal differentiation of human-environment relations, to the so-called quantitative and theoretical revolution of spatial science (which made more rigorous the accurate description of material geographies), to the radical redirection of spatial science in Marxist, feminist and other 'critical' human geographies (seeking not simply more accurate description but 'emancipatory' explanation, understanding and practice). The objects of analysis and explanation in all these phases are the concrete material forms of empirically 'real' geographies seen as outcomes of what are typically presumed to be influential but usually non-spatial processes, such as those arising from the exigencies of societal cohesion, ecological adaptation, capitalist accumulation or the maintenance of patriarchal power.

A subjectivist or idealist geography has often arisen as a counterfoil to excessive objectivism, materialism and scientism, not unlike what has occurred in many other disciplines. Within the field, these subjectivist approaches, drawing deeply from humanistic, existential, phenomenological and cognitive or psycho-behavioural traditions, have tended to take on connotative aspects of the subaltern against the materialist hegemon. Except on very rare occasions, the material geographies remain the primary objects of explanation, understanding or interpretation, but the implied causality is seen as primarily ideational and subjective, formed in what are primarily 'imagined' geographies rather than arising directly from objective material social (or other non-spatial) processes.

To borrow, with some modification, from Lefebvre, the first mode is fixed on the real 'spaces of representation' while the second is focused on the imagined 'representations of space'. It is easy to say (as most geographers do) that both are important and should be combined in good geographical analysis, but too often in the history of geographic thought, subjectivism and objectivism, imagined and real geographies, have been placed in rigid opposition, especially when couched in extreme or essentialist forms of idealism or materialism. Philosophical and methodological critique in modern geography has tended to be dominated by this categorical opposition or by the struggles for intrabinary primacy between specific forms of materialist and ideational analysis, fostering a continuation of this divisive split and either/or competition.

Here is one point where the earlier discussion of binary orderings of difference and the critique of modernist identity politics becomes particularly relevant. In the new cultural politics of difference, the aim is neither simply to assert dominance of the subaltern over the hegemon in a rigidly maintained bipolar order, nor even to foster some specified combination of opposing traits and traditions. It is to break down and disorder the binary itself, to reject the simple structure of closed dualisms through a (sympathetic) deconstruction and reconstitution that allows for radical openness, flexibility and multiplicity. The key step is to recognize and occupy new and alternative geographies – a 'thirdspace' of political choice – different but not detached entirely from the geographies defined by the original binary oppositions between and within objectivism and subjectivism.

For Lefebvre, this alternative geography is the socially produced spatiality wherein the directly experienced spaces of representation and the conceptual representations of space can strategically interrelate in the lived (*vécu*) contextuality of *spatial praxis*. Lefebvre always remained a Marxist and a materialist first, but his counter-hegemonic spatial praxis sought to recombine materialism and idealism and to reach beyond the parochialisms and illusions of both. He was able to achieve this recombinative simultaneity only by opening up a new terrain, by finding new sites for active resistance and critical dialogue.[22] A similar thirdspace of political choice also figures prominently in Foucault's heterotopias, 'formed in the very founding of society' as 'something like counter-sites, a kind of effectively enacted utopia in which the real sites, all the other real sites that can be found within the culture, are simultaneously represented, contested, and inverted'. Heterotopias, combining the real and imagined, are the 'space in which we live, which draws us out of ourselves, in which the erosion of our lives, our time and our history occurs, the space that claws and gnaws at us.'[23]

We contend that the spatiality of bell hooks and others involved in the postmodern discourse on the new cultural politics can best be located and understood in this thirdspace of political choice – and inversely, that the explorers of this different spatiality cannot be appropriately seen and understood by those confined by the more traditional spatialities of modernism, whether trained as geographers or not. Hence our call for the development of *postmodern geographies* as a radical standpoint, perspective and positioning from which we can begin the process of re-visioning spatiality in a contemporary world where all real geographies are imagined and all imagined geographies are real.

Our final argument has to do with a more specific counter-hegemonic project that may be particularly suited to the marginal position occupied by the discipline of geography in the twentieth-century intellectual division of labour. The discipline can be described as having been doubly marginalized over most of the past century, first in conjunction with the modern social sciences and humanities (and for physical geography, within the physical sciences as well), with geography dwelling intellectually in an often forgotten periphery of utilitarian fact-gathering and map-making; and second, in the historical development of specifically social and political theory and philosophy, where the ontological and epistemological hegemony of history over geography, the temporal over the spatial, as well as the discipline of geography's own parochial spatialisms, further intensified the discipline's isolation and peripheralness. We suggest here that this peripheral positioning can now be used as a site of opportunity, another place of radical openness where new alternatives can be imagined and effectively practised by consciously and strategically disordering difference and choosing marginality.

We thus raise again the issue of *historicism* and hegemonic historiography noted by Spivak, Said and Hebdige. This privileging of the 'making' of history and the critical historical imagination over the 'making' of geographies and what should be the equally revealing and emancipatory power of the critical geographical imagination continues largely unquestioned and unacknowledged even among many postmodern cultural critics and geographers. The power engrained in the real and imagined spatiality of social life is recognized and creatively explored, but such exploration too often remains auxiliary to the superordinate significance of historical understanding and interpretation.

In identifying this alleged epistemological hegemony of historicism, it is difficult to avoid appearing simplistically antagonistic to the rich emancipatory insights of critical historiography: to be somehow 'against' time and history and 'for' space and geography. This is not the intention of the contemporary critique of historicism. Instead, the aim is to deconstruct and reconstitute the geographical and historical imaginations in a critical 'trialectic' that revolves around the problematic interrelations between historicity, spatiality and sociality; or, more concretely and consciously political, the (social) making of histories, the (social) production of human geographies, and the (spatio-temporal) constitution of social practices and relations. Here again, the assertively spatialized discourse on the new cultural politics of difference vividly enters the debate, providing another way to clarify and redefine its meaning and intent: that from whatever disciplinary perspective we come from, to be critical thinkers we must all be historians, geographers and social analysts.

The critique of historicism can appropriately be seen, in this new light, as a critique of the constraining effects of an intellectually hegemonic historiography and an intellectually subordinated geography. Its tactics are not those of categorical inversion, constructing an essentialist spatialism in historicism's place, but of decentring the hegemonic subject, of disordering differences and creating a new terrain of reassembly, a common ground of multidisciplinarity, where we can be historical and spatial and critically social simultaneously and without a priori privileging of one or another viewpoint. This too is a space of openness, an 'inclusive space where we recover ourselves', where one can be seen and be in contact with other sites of radical subjectivity and counter-hegemonic resistance. As bell hooks has said: Marginality is the space of resistance. Enter that space. Let us meet there.

NOTES

1 After a powerful critique of conventional postmodern discourse for its white male exclusivity, especially its separation of the 'politics of difference' from the 'politics of racism', bell hooks reasserts her radical postmodernism with no regrets: 'Radical postmodernism calls attention to those shared sensibilities which cross boundaries of class, gender, race, etc., that could be fertile ground for the construction of empathy – ties that would promote recognition of common commitments, and serve as a base for solidarity and coalition. . . . To change the exclusionary practice of postmodern discourse is to enact a postmodernism of resistance' (hooks 1990: 27, 30).

2 Our shared postmodernism is tinged with significant differences which we hope to preserve in our presentation. Identifying these differences in advance can be misleading, but in a profile across the multiple forms and foci of resistance noted by Cornel West (exterminism, empire, class, race, gender, sexual orientation, age,

nation, nature, and region), one co-author's postmodern feminism foregrounds (without privileging) gender and sexual practice and the other's postmodern geographical Marxism foregrounds (also without privileging) class and region.

3 Following Foucault, subjection, domination and exploitation define three broad types of struggle against hegemonic power. The first (least well-developed and understood) reflects the double meaning of the word 'subject': subject to someone else by control and dependence vs. tied to one's own identity by conscious self-knowledge. Struggles against subjection are thus struggles over subjectivity and submission, over who defines the subject, individually as well as collectively. Foucault argues that 'nowadays, the struggle against the forms of subjection – against the submission of subjectivity – is becoming more and more important', although he hastens to add that struggles against forms of domination (ethnic, social, religious) and exploitation (against alienation, the social and spatial relations that separate individuals from what they produce) continue to prevail. We combine these three forms of struggle in our definition of cultural politics and raise the increasing importance of struggles over subjectivity and submission to a key position in defining the *new* cultural politics of difference. See Foucault (1982).

4 This need to emphasize geography and spatiality in what has often been an almost exclusively historical discourse and theorization of uneven development is a central theme in Edward W. Soja, *Postmodern Geographies: The Reassertion of Space in Critical Social Theory*, London, Verso, 1989. We also will argue throughout this chapter that the reassertion of a spatial perspective, discourse and mode of theorization has been a distinctive feature of the new cultural politics of difference.

5 These observations must not be taken as privileging class struggle as paradigmatic or as a model which other forms of modernist resistance mimicked. Such privileging is largely the result of Eurocentric and masculinist/phallocentric analyses of social reality, opposition and resistance as they developed (and became hegemonic as 'Marxist') in the nineteenth and twentieth centuries. It is precisely this essentialist privileging of the working class, the radical (male) intellectual, and later the Party as *the* agents 'making history' – and the attendant deployment of Marxism as an all-inclusive, self-sufficient grand narrative of social revolution – that radical post-modernism and the new identity politics (as well, we might add, as some nineteenth-century feminisms and anti-racisms) have worked to deconstruct. Even drawing parallels and analogies must be done with care, for the emphasis on exploitation that is so central to class struggle is significantly different from the emphasis on subjection and domination which motivates most other forms of cultural politics.

6 The generalized bipolarity of the relation between hegemon and subaltern can be expressed in more emphatically spatial terms as the relation between core (or centre) and periphery (or margin). In both cases, the binary ordering of difference is socially constructed (not natural or transhistorical) and arises from the workings of power in a complex process of geohistorically uneven development. The homology (rather than mere analogy) of these two expressions of binary ordering is another important stimulus to the spatialization of the cultural politics of difference.

7 Here is where the universalist politics of liberal humanism (and its attendant individualism) frequently complicates the theory and praxis of radical identity politics by providing an alluring, all-embracing, alternative viewpoint. This categorically simplified choice – occupy only one 'radical' channel or 'liberally' choose them all, especially when backed by the similar categorical binarism of 'if you are not our friend, you are our enemy' – imposes a rigid logic which intensifies the need for an 'epistemological police' to guard boundaries and punish deviation, especially for those who seek a different set of choices.

8 Further intensifying this tendency to fragmentation (and expanding the role of the epistemological police) is an intrabinary competition for primacy, driven by the same 'urge to unity' that pervades modernist identity politics. Here we find the effort to exclude, to separate the pure from the impure, the deviation from the essence: a demarcation of categorical distinctions between inside and outside that actively represses difference and otherness. See the work of Diana Fuss (1991).

9 It is important to note the emergence of an increasingly powerful reactionary form of postmodern politics that also engages in a 'disordering of difference' and has learned effectively how to create political advantage from fragmentation, ruptures, discontinuities and the apparent disorganization of capitalist (and formerly socialist) political economies and cultures. In many ways, the hegemon has learned to adapt to the conditions of postmodernity more rapidly and creatively than the subaltern, especially when the latter remains frozen between modernist and postmodernist radical subjectivities.

10 One might argue as well that a similar tendency can be found in our critique of modernist identity politics. We hope that we have not been so categorically dismissive of modernism or to have homogenized modernist identity politics to the same degree that is present in the contemporary radical backlash against postmodernism. We do recognize, however, that there is a widespread tendency to be simplistically and categorically anti-modernist (more specifically, anti-Marxist and anti-feminist) among many who proclaim themselves to be postmodern. For an attempt to clarify these many positionings, see Edward Soja, 'Postmodern geographies and the critique of historicism', in *Postmodern Contentions: Epochs, Politics, Space*, ed. J. P. Jones, W. Natter and T. Schatzki, New York, Guildford Press (1993).

11 All the other quoted references to bell hooks are taken from 'Choosing the margin as a space of radical openness', in *Yearnings*, 145–53.

12 The full quote from Foucault comes from 'The Eye of Power', reprinted in *Power/Knowledge: Selected Interviews and Other Writings 1972–1977*, ed. C. Gordon, New York, Pantheon, 1980, p. 149. Foucault writes: 'A whole history remains to be written of *spaces* – which would at the same time be the history of *powers* (both terms in the plural) – from the great strategies of geopolitics to the little tactics of the habitat.' The same quote reappears as the lead-in to Daphne Spain, *Gendered Spaces*, Chapel Hill, University of North Carolina Press, 1992.

13 See, for example, *The Book of the City of Ladies*, written by Christine de Pizan (a Frenchwoman born in Venice in 1364), trans. E. J. Richards, New York, Persea Books, 1982. For a tracing of the feminist critique of urbanism and the 'gendering' of the city (and of urban and regional planning theory) in Western capitalist societies, see Hooper (1992).

14 A sampling of this literature would include Hayden and Wright (1976); the special issue of *Signs* on 'Women and the American City' 5 (1980); Werkele, Peterson and Morley (1980); Hayden (1982); Ardener (1981); Mazey and Lee (1983); Matrix Collective (1984); Women and Geography Study Group of the Institute of British Geographers (1984).

15 In just the past year, at least four notable (and very different) books have appeared from US publishers exploring the spaces that difference makes. Daphne Spain (1982), in *Gendered Spaces*, critically synthesizes the modernist feminist literature and develops further the strategy of degendering spaces in dwellings, schools and workplaces. *Sexuality and Space* (Colombia 1992) foregrounds sexuality and the body in the postmodern politics of space; while Celeste Olalquiaga (1992) infuses the 'culturescapes' and 'cyberspaces' of the city with wildly postmodern (and postcolonial) sensibilities. Finally, Leslie Kanes Weisman (1992) explores 'bodyscape as landscape' and urban spatiality as a 'patriarchal symbolic universe'.

16 See Deutsche (1988; 1991), Morris (1988a; 1988b), Wilson (1991).

17 See Young (1990), Fraser (1990), Haraway (1991).

18 As Fuss makes clear, there will always be significant tensions and dangers in choosing marginality as a space of resistance, not the least of which is the temptation towards 'romancing the margins', assuming the margins to be the sole 'authentic' location for the voices of truth and justice. That the dangers exist even after deconstruction must also be recognized. For a similar reading of the life and death of Henri Lefebvre as a centred peripheral, see Soja (1991).

19 In her own act of displacement Spivak writes: 'I find the demand on me to be marginal always amusing. . . . I am tired of dining out on being an exile because that has a long tradition and it is not one I want to identify myself with. . . . In a certain sense, I think there is nothing that is central. The centre is always constituted in terms of its own marginality. However . . . certain peoples have always been asked to cathect the margins so that others can be defined as central. . . . In that situation the only strategic thing to do is to absolutely present oneself at the center' (Spivak 1990: 41).

20 Dick Hebdige, introduction to a special issue on 'Subjects in Space', *New Formations* 11, 1990, vi–vii. The quoted references are to Michel Foucault (1980a, 1986). For a further elaboration upon these key Foucauldian writings, see Soja (1990).

21 Michael Keith and Steve Pile, 'Communities of Resistance: Geography and a New Politics of Identity', prospectus circulated to participants in a session organized by the Social and Cultural Study Group of the Institute of British Geographers for the Annual Meeting, Swansea, January 1992.

22 Lefebvre first named this new terrain 'everyday life' and later 'the urban'. In both cases, but especially with his urban 'approximation', he was often misunderstood by even his most enthusiastic supporters. He tried for many years to explain that his use of the urban did not refer literally only to the material forms and geographies of the city or to conventional visions of urbanism as a way of life, but to another meaning of urban: simultaneously real and representable, imagined and represented, a more broadly defined, encompassing and combinatorial social spatiality which he described as a grounded metaphor, a concrete abstraction, a social heiroglyph. See Lefebvre (1991).

23 From his *Des espaces autres*, translated as 'Of Other Spaces'. The quotations can be found in Soja (1990).

REFERENCES

Ardener, S. (ed.) (1981) *Women and Space: Ground Rules and Social Maps*, New York: St Martin's Press.

Cocks, J. (1989) *The Oppositional Imagination: Feminism, Critique and Political Theory*, London: Routledge, 20.

Colomina, B. (ed.) (1992) *Sexuality and Space*, New York: Princeton Architectural Press.

de Pizan, C. (1982) *The Book of the City of Ladies*, New York: Persea Books.

Deutsche, R. (1988) 'Uneven development: public art in New York City', *October* 47: 3–52.

——— (1991) 'Alternative space', in B. Wallis (ed.) *If You Lived Here: the City in Art, Theory and Social Activism*, Seattle: Bay Press, 45–66.

Foucault, M. (1980a) 'Questions of geography', in *Power/Knowledge: Selected Interviews and Other Writings 1972–1977* ed. C. Gordon, New York: Pantheon, 63–77.

——— (1980b) 'The eye of power' in *Power/Knowledge: Selected Interviews and Other Writings 1972–1977* ed. C. Gordon, New York: Pantheon.

——— (1982) 'Afterward: the subject and power', in M. C. Dreyfus and P. Rabinow (eds) *Michel Foucault: Beyond Structuralism and Hermeneutics*, Chicago: University of Chicago Press, 208–26.

——— (1986) 'Of other spaces', *Diacritics* 16: 22–7.

Fraser, N. (1990) 'Rethinking the public sphere: a contribution to the critique of actually existing democracy', *Social Text* 25/26.

Fuss, D. (1991) 'Inside/out', in D. Fuss (ed.) *Inside/Out: Lesbian Theories, Gay Theories*, New York: Routledge.

Haraway, D. (1991) *Simians, Cyborgs and Women: the Reconstruction of Nature*, New York: Routledge.

Hayden, D. (1981) *The Grand Domestic Revolution: a History of Feminist Designs for American Homes, Neighbours and Cities*, Cambridge Mass.: MIT Press.

Hayden, D. and Wright, G. (1976) 'Architecture and urban planning', *Signs: a Journal of Women in Culture and Society* 1: 923–33.

hooks, b. (1990) *Yearnings: Race, Gender and Cultural Politics*, Boston: South End Press.

——— (1992) *Black Looks: Race and Representation*, Boston: South End Press.

Hooper, B. (1992) 'Split at the roots: a critique of the philosophical and political sources of modern planning doctrine', *Frontiers: a Journal of Women Studies* 13.

Lefebvre, H. (1991) *The Production of Space*, Cambridge, Mass.: Blackwell.

Matrix, Collective (1984) *Making Space: Women and the Man-made Environment*, London: Pluto Press.

Mazey, M. E. and Lee, D. R. (eds) (1983) *Her Space, Her Place: a Geography of Women*, Washington, DC: Association of American Geographers.

Minh-ha, T. T. (1991) *When the Moon Waxes Red: Representation, Gender and Cultural Politics*, New York: Routledge.

Morris, M. (1988a) 'At Henry Parkes Motel', *Cultural Studies* 2(1): 1–47.

——— (1988b) 'Things to do with shopping centres', in S. Sheridan (ed.) *Grafts: Feminist Cultural Criticism*, London: Verso.

Olalquiaga, C. (1992) *Megalopolis: Contemporary Cultural Sensibilities*, Minneapolis: University of Minnesota Press.

Parmar, P. (1990) 'Black feminism: the politics of articulation', in J. Rutherford (ed.) *Identity: Community, Culture, Difference*, London: Lawrence & Wishart, 101–26.

Said, E. (1979) *Orientalism*, New York: Vintage.

——— (1985) 'Orientalism reconsidered', in F. Barker *et al*, (eds) *Europe and Its Others*, Colchester: University of Essex.

Soja, E. (1989) *Postmodern Geographies: the Reassertion of Space in Critical Social Theory*, London: Verso.

——— (1990) 'Heterotopologies: a remembrance of other spaces in the Citadel-LA', *Strategies: a Journal of Theory, Culture and Politics* 3.

——— (1991) 'Henri Lefebvre, 1901–1991', *Environment and Planning D: Society and Space* 9.

——— (1993) 'Postmodern geographies and the critique of historicism', in J. P. Jones, W. Natter and T. Schatzki (eds) *Reassessing Modernity and Postmodernity*, New York: Guildford Press.

Spain, D. (1992) *Gendered Spaces*, Chapel Hill: University of North Carolina Press.

Spivak, G. C. (1988a) 'Subaltern studies: deconstructing historiography', in R. Guha and G. C. Spivak (eds) *Selected Subaltern Studies*, New York: Oxford University Press.

——— (1988b) *In Other Worlds: Essay in Cultural Politics*, New York: Routledge.

——— (1990) *The Post-Colonial Critic: Interviews, Strategies, Dialogues*, New York: Routledge.

Trinh (1989) *Women, Native, Other: Writing Postcoloniality and Feminism*, Bloomington: Indiana University Press.

Weisman, L. K. (1992) *Discrimination by Design: a Feminist Critique of the Man-made Environment*, Urbana and Chicago: University of Illinois Press.

Werkele, G., Peterson, R. and Morley, D. (1980) *New Space for Women*, Boulder, Col.: Westview Press.

West, C. (1990) 'The new cultural politics of difference', in R. Ferguson, M. Gever, T. T. Minh-ha and C. West (eds) *Out There: Marginalization and Contemporary Cultures*, Cambridge, Mass.: MIT Press; New York: The New Museum of Contemporary Art.

Wilson, E. (1991) *The Sphinx and the City: Urban Life, the Control of Disorder, and Women*, London: Virago Press.

Women and Geography Study Group of Institute of British Geographers (ed.) (1984) *Geography and Gender: an Introduction to Feminist Geography*, London: Hutchinson.

Young, I. M. (1990) 'The ideal of community and the politics of difference', in L. J. Nicholson (ed) *Feminism/Postmodernism*, London and New York: Routledge, 300–23.

11

QUANTUM PHILOSOPHY, IMPOSSIBLE GEOGRAPHIES AND A FEW SMALL POINTS ABOUT LIFE, LIBERTY AND THE PURSUIT OF SEX (ALL IN THE NAME OF DEMOCRACY)[1]

Sue Golding

> Ye gods! annihilate but space and time
> And make two lovers happy.
> (Alexander Pope 1727)

Before I begin this chapter in earnest, I would like to give a general idea of the fairly severe (and rather strange) positions being developed here. There are three. First, I'm raising the question of an impossible spatiality: impossible not because it does not exist, but because it exists and does not exist exactly at the same time. This is what I shall name a 'radical geography', which, still under point number one, I'd like to suggest, can best be understood by relying on, at least in part, what I would like to introduce here as 'quantum philosophy' (having grown tired of adjectives like 'post-' or 'post the x': post-structuralism, postmoderism, post-liberalism, post-feminism and the like; terms which seem to do little more than focus the argument on exactly what is being criticized, attempting to be superseded or, worse, put forward 'as if' it were always already superseded and ought to be so). Point number two: I want to argue that, apart from ivory tower sport, one reason to excavate or better, acknowledge, a 'quantum philosophy' is that it points to what I think is an implicit *ethical* (and political) assumption in the conceptualizations of what it 'is' to be a person as well as what it 'ought' to be within the kind of political theory or political philosophy that has now come to be known as discursive theory, replete with its logics of contingency, and so forth.

Let's put this slightly differently: in as much as every political theory always carries with it a conception of 'what is human nature', I would like to venture that, likewise, there is indeed a specific concept operating in discursive theory. And I would like to venture, furthermore, that that conception is one which places masochism, and let's be even more explicit and say *sexual* masochism, as a fundamental, radically hegemonic 'grounding', as it were, not only for this social animal we call human, but for this entity we call society. This point of departure has very specific ramifications on a number of levels, not the least of which – having acknowledged, among other radical dimensions, this arena called the sexual – has something to do with what can be called 'queer politics'. As we will see, it's a kind of queerness which draws upon, among other things, the whole lesbian, gay and pervert movement, replete with its blood, sweat, pleasures and death.

Third and finally, I want to say that points one and two set up, at least in part, a way to get at the whole question of social 'progress' itself; a sociality that I would like to suggest has, heretofore, *implicitly* relied on the importance of the urban or urban-ness, i.e. the anomic yet multiple, heterogeneity of the city (or perhaps, rather cities). I want, instead, to foreground the question of, indeed the necessity of, urban-ness, in order to suggest that the *urban itself, and the kinds of cityscapes to which it points, is a fundamental requisite for radical pluralism*. Indeed, I will go further, and suggest that the kind of 'urban' politics being recognized here not only belies the very sense of a radical pluralism, but in its wake dislodges and undoes the very notion of 'community', community politics and the like, without at the same time capitulating to, or privileging, the individual over and against (or apart from) society. At this juncture, then, we come full circle (or perhaps it's a kind of spiral) to the very understanding of democracy itself, a specific kind of democracy existing in the epoch of – for lack of a better way of putting it – technology; or, to accept the usual phraseology, as existing in and expressing what is often called the 'information age'.

A *statement* by Nietzsche:[2]

> '11 Can an *ass* be tragic? To perish under a burden one can neither bear nor throw off? The case of the philosopher.'

To this I now add an a, b, and a c:

> 11a Can a *space* be impossible? To emerge in the geography of its 'not-space', in the in-between of its potentiality and its erasure? The case of radical democratic possibility.
>
> 11b Can a *transgression* of the body-flesh have a point? To play past corruptible infinitude and yet be bound, mercilessly, in the web of time? The case of the (discursive) ethical-political.
>
> 11c Can a *cell*, biological or otherwise, have a memory? To be immunized at birth against a disease at age 60? The case of the urban-city as anti-body; the case, that is to say, against community.

The three addendum to the tragedy of the ass: first, the (not-space)-space of space and, with it, time; second, the masochistic ethicality of corruptible infinitude; third, the memory of an un-dead trace operating 'as if' it were always-already still there, disfiguring the homogeneity of a totalized body, while at the same time, preserving or extending (or maybe a better word, inventing) its existence. In a very direct sense, these are concepts well worn in our contemporary philosophic discourse. But today I want to recover them in the fullest sense of that verb 'to recover': that is, to bring to them the surface, to hide their faults with a likeable pattern, to make them 'as good as new', but to do so by doubling the 'membering', by *re*-(re)membering the 'what', the 'that', the 'how'. That is, by doubling the membering of the techne – and not at all the 'why' – of the distance between a re-upholstered concept and the resemblance of its other, a kind of fabric stitched between its present form and its past, be it public or private or underground or fictional or any combination in between. And then we shall *re*-cover them; that is, submerge them, sacrifice them, or shall I say at the risk of sounding even more poetic, eat and digest these not-so-innocent little lambs as the (un)thinkable 'grounding' for the possible.

11a THE SPACE OF THE NOT-SPACE (OR WHAT'S SO INTERESTING ABOUT QUANTUM PHILOSOPHY)

My fears. I will begin with my fears. First fear: when you hear the concept of 'space' (or rather, the space of the 'not-space'), you might immediately think I am referring only, or predominantly, or even at all to a visual presencing; that is, to a kind of positivity or direct positionality, i.e., an exacting geography, a marked or bounded arena, of space meant in some way to 'ground', as it were, the concept of time. Or conversely, fear number two: you might think we're grappling here with the text of an 'empty space', à la a Lefort or a Furet – not that I am at all opposed, necessarily, to their conceptualizing of democracy as 'an empty place, impossible to occupy' whereby it 'conjoins two apparently contradictory principles: one that power comes from the people, the other that it is the power of nobody'.[3] Irrespective of one's position on theirs, this so-called 'possible', this 'not-space' spacing cannot be pocketed in that way.

Next fear: perhaps this 'imaginary' geography could be perceived as a kind of neo-Bachelardian return to the poetics of space, a kind of (post) 'modern day phenomenology of the soul, indispensable without mind', as it is itself phrased (sans the (post)) in Bachelard's well-known *Poetics*.[4]

Continuing fear: that this recovery be understood as an attempt to resurrect the ontological status of a 'Being-in-the-world', one that includes 'de-severance', a 'dis-location', a kind of 'directionality', and hence as something which – with Heidegger (literally) underscoring in para. 23, of *Being and Time* – posits Dasein, in one sense, as that which is '*essentially . . . spatial*'.[5] (Parenthetical remark: on the other hand, it is certainly not by accident that this radical geography, i.e., this 'between of space', is indeed being re-covered; and there is a certain, very real, hat-tipping toward (but against) Heidegger's insistence on the notion of 'giving space').

Last conscious fear: a recuperation of the 'space' of the 'not-space' might simply be taken as a metaphor, or worse, simply as a more cumbersome way of 'saying' or 'containing' or even 'hinting' at the notion of the lack, at least in its self-reflexive sense.[6] We will return to this point soon enough, but let's mark it for now with a nod toward Lacoue-Labarthe.[7]

The kind of 'between', the kind of 'space of the not-space', the kind of 'radical geography' towards which I want to direct our attention has more to do with Wittgenstein's playful remarks on the law, and more specifically, on (but against) the 'law of excluded middle'. Indeed, as will be noted momentarily, it has also to do with what I shall call 'quantum philosophy'.

In Part V of his *Mathematics*, Wittgenstein sets up the problem like this:

> 10. When someone hammers away at us with the law of excluded middle as something which cannot be gainsaid [i.e., questioned], it is clear that there is something wrong with his question.
>
> When someone sets up the law of excluded middle [so that there cannot possibly be anything that exists between 'p' and 'not-p' or outside their unity], he is as it were putting two pictures before us to choose from, and saying that one must correspond to the fact. But what if it is questionable whether the pictures can be applied here?
>
> And if you say that the infinite expansion must contain the pattern ϕ or not contain it, you are so to speak shewing us the picture of an unsurveyable series reaching into the distance.
>
> But what if the picture began to flicker in the far distance? [How does one account for it then?]
>
> (Wittgenstein 1983: 268)

Can we think of, in other words, a different way to come at the whole question of the limit (or of division), in this case, between the 'that' and its other, without setting up the usual two camps, i.e., the usual binaries, (like insides and outsides, or even, for that matter, male and female) with no slippage between a so-called positivity and, inversely, its so-called negativity? Moreover, can we do this without resorting to some infantile form of morality, wherein the 'that' which is 'real' or 'good' corresponds to an objective 'truth'; the other, to its subjective 'lie'? And is it possible to do this, without resorting, in the end, to an originating Archimedean point existing above or beyond the real (let's say, for example, like God, or capital 'T' Truth, or Universal Ontology) or without reinventing some transcendental monad arising from a dialectical synthesis, historical or otherwise. That is, is it possible to come at the difficult question of 'what ought to be', not to mention, 'what is', without lapsing into pre-given truths or, conversely, into what is often understood as the trap of relativity (i.e., as implying a kind of ethic of whatever is created, 'as such', is 'good')?

It seems to me that this is precisely what is at stake, at least methodologically speaking, in discursive theory. Interestingly enough, I would say it is also what is at stake in quantum physics. The connection runs something like this: If truth is no longer pure 'objectivity'; if, in other words, it is no longer to be understood simply as a mimetic repetition of sameness, i.e., an actual point-for-point mirrored reflection, an equivalence, a 1:1 ratio; if this kind of logic no longer suffices to get at a fuller explanation of the picture 'flickering in the distance'; if, that is to say, finally, we have indeed moved into the nether regions 'beyond' Good and Evil (since it is now possible to think the entirety of the field as no longer exhausted by p/(not-p)), then we have to recognize the possible existence of an 'excess' to the Other, a possible not-Other – something that Gramsci calls 'the disturbing element' of politics, and with it, the will; something that in quantum physics is known as 'the elsewhere'. This excess of the Other is a 'quasi-something' (or even a 'quasi-nothing'), a kind of quasi-concept at any rate, negotiating the very cohesiveness of this infinite field called reality, and doing so without in any way being posited 'above' or 'a priori' the event, or even as its synthetically dialectic expression. There is no teleological unfolding toward anything, ethical or otherwise. We have here, instead, a quasi-entity, a strange kind of 'excess' (strange because it exists neither within nor without reality); we have here precisely an impossible spatiality, a radical geography, existing and not existing exactly at the same *time*.

In philosophy, we have tried to describe this in various ways calling it something like 'the truth of fiction' or 'thought from outside' (Foucault/Blanchot) or the logic of the gramme or trace (Derrida), i.e., a constitutive 'that' which is not recuperable, something not absorbed into the objective field of the Law of Logos, but able at the same time to 'contour' it, without itself becoming some kind of limit or negation or void. This kind of non-representational 'something' is understood, strictly speaking, as the temporal moment, a kind of aesthetic *dis*-location, not reducible to the fixities of space (Laclau). Yet, here, I would venture to say, that however powerful and important this kind of heterogeneic logic or multiplicitous contingency has been (and remains), we are still stuck in two (or at best, three) dimensions. And this remains problematic, not the least because we still cannot grapple with the entirety of the spectrum (or spectra) of life in all of its impure and profane discrepancies. Perhaps we never will.

In quantum physics, however, another leap could be made, and indeed, already has been. For here one finds this impossible 'something' understood not just in terms of the corruptible (i.e., non-absolute) temporal moment *per se*, but as a kind of spatial 'patch' (perhaps as a kind of nodal point, perhaps as a kind of line (Or is a line just an extended point? Or is a line a series of discrete points? Or is it a wave? Or a curve? Or all of the above, or none of them?), in any case a kind of space impossible to grasp and yet able to be specified by at least four, *completely relative*, co-ordinates: length, depth, width and now, also, speed. Einstein was one of the first, of course, to recognize this in his early formulation of the theory of relativity,[8] summarizing the problem in terms of the now infamous equation, $E = MC^2$.

To put this slightly differently, we have now before us not only a fluid (and yet discrete) concept of time, but also a dynamic concept of space, a concept, to be even more precise, that debunks any notion of an 'eternally infinite' spatiality, while simultaneously refusing its uniqueness as if only geographically 'singular', that is to say, 'closed', 'totalized', 'homogeneous' and therewith, always-already, 'fixed'. We have here, in other words, 'space-time', an imaginary, but real, and utterly *dynamic*, fourth dimension. A kind of infinite finiteness (already a strange concept) contouring the universe, making the 'that which stretches out into eternity' always fractured by a limit, itself 'unique', relative and infinitely contingent. But we have here, also, another strange, almost completely mad, concept: we have here a finite inifinitude. The universe, which heretofore was understood as having existed forever a priori the beginning of time and forever after it, now, given Einstein's theory of relativity and its ironic play with and against the fourth dimension of space-time, 'implie[s] that the universe must have a beginning and possibly, an end'.[9]

'God is dead', as Nietzsche solemnly intoned, or perhaps, better put by Bataille, 'God is a whore', a corruptible and decaying infinitude. Suddenly we have a whole series of weird quasi-concepts erupting out of this: like having 'beginnings' with no originary point, like having endings with no limit. Usual little concepts like the notion of a 'first' or that of a given 'law', lose their sense of hierarchy or permanence; and in their having become a beginning or having remained 'steadfast', they take on the garments of discursivity and contingency at their very core. Like the stars, let's say, they become nodal points with boiling centres held together by 'the that' which is its not; limited only by the speed of infinitely finite light waves (i.e., eternally discrete time) and by the very curve of a multi-dimensional elsewhere.

All this is very interesting, but the question must be put: What on earth does this recuperation (or re-invention) of a radical geography, i.e., of a fourth dimensional space-time, a quantum philosophy, have to do with democratic possibility? Moreover, what does this have to do with sexual masochism or queer politics, not to mention the question of the city? What does it have to do with *us*?

11b CORRUPTIBLE INFINITUDES AND THE 'WHAT OUGHT TO BE'

There are several ways to approach these questions. I will attempt only one. And that is to take seriously the logic of contingency (and with it, multiplicity; indeed, also, the politics of hegemony) as the primary and irreducible relation to any truth-game, where the words 'primary' and 'irreducible' are understood here in a discursive and not hierarchical sense. If we push the implications of this logic as far as they will go – or at least further than what has been done up to now – a funny thing around ethics, political strategies and democratic possibilities begins to surface. Foucault characterized this odd little thing in one word: transgression. Bataille called it: the sacred. I want to call it: queer. And by re-inventing (or re-covering) this word, I mean to bring with it the whole damn business of sweat and blood and pleasure and death; that is to say, all that is implied around transgressive, and sacred sex. Let's clarify exactly what is at stake.

It begins on and against the 'ground' of space-time, the ground, that is to say, of both irreducible multiplicity and the constance of instability, the ground that presents the 'given', i.e., the Law, as itself a rule no longer static nor dialectic nor representational nor lack. It is a law both of all of these things and of none of them; one that is private and public and underground and fictional, and for all that (or perhaps because of it), forces us to stand before this Law, to be compelled by it, and yes, even to shape it ourselves, precisely along the same axes, along the same doublings, entanglements and denegations in all their varied complexities. Wittgenstein casts it this way, going so far as to call it a 'spell':

> If a blind man were to ask me 'Have you got two hands?' I should not make sure by looking. If I were to have any doubt of it, then I don't know why I should trust my eyes. For why shouldn't I test my *eyes* by looking to find out whether I see my two hands? *What* is to be tested by *what*? . . .[10]
>
> The difficult thing here is not to dig down to the ground; no, it is to recognize the ground that lies before us as ground. For the ground keeps giving us the illusory image of a greater depth, and when we seek to reach this, we keep on finding ourselves on the old level. Our disease is one of wanting to explain. [But] 'One you have got hold of the rule, you have the route traced for you.'[11]
>
> [And yet the question becomes:] What is it that compels me? – the expression of the rule? – Yes, once I have been educated in this way. But can I say it compels me to follow it? Yes: if here one thinks of the rule, not as a line that I trace, but rather as a spell that holds us in thrall.[12]

Now, this is slightly different (actually, completely at odds with) liberalist assumptions around the Law and one's relation to it. For in the liberalist case – allowing for the moment the right to lump so broadly into one category the whole history of liberalism with its democratic variants – the Law is Reason; and 'man' stands in relation to this reason; indeed can become it; can become, that is to say, 'master': master of one's home, master of one's property, master of one's self-hood, and with it, master of the body and all its corruptible pleasures. In liberalism we have here, in other words, the notion of Law as an unrelenting master, and as such, as both ground and horizon – wherein reason-as-logos-as-law is both a fully inscribed 'is' as well as an 'ought'; i.e., as 'a something' to be guided by and 'a something' to become, over and against the web of bodily decay and the temptations of the flesh. I would dare say that the very basis of civil rights in a liberal democracy is based on this one, utterly primordial fact of (a closed notion of) mastery, power and law.

Does it seem odd to you, then, that so many progressive liberation movements of this century – black, women, civil, communist, socialist, environmentalist, anarchist – even those within the music industry, be it punk, house, jazz or rock-'n'-roll – have always included as a basic demand (at least in the beginnings of the movement so named), the right of *sexual* 'freedom', the right that is, to go 'beyond', i.e., to 'transgress' the master-father-status-quo? That is to say, and to be absolutely clear about it, to include not just 'any' kind of transgression, but one that involves the erotic pleasures (and dangers) of the genitals? This should seem no more 'odd' than the fact that in most cases that very demand, albeit as 'transgressive' and as anti-establishment as it could get, often and of necessity, became fully absorbed into (or crushed under) the Law. . . . With the possible exception, I hasten to add, of the pervert moment, euphemistically (or otherwise) named: gay.

Why should being 'queer', that to say, being a sexual pervert – and, in the context I want to emphasize here today, that of the lesbian 'outlaw' – be any different? The short answer is that it's not. It's particularly not any different, and certainly will do no better than become 'absorbed', if the 'transgression' of the Law is based either on a notion of Law as a fixed horizon/ground; or on the notion of 'transgression' as simply a 'not' or 'anti-' the existing x, y, and z's of life. The longer answer is that it might possibly be, and given the history/present of gay movement politics, it seems, very well, that it is.

Let's look at it like this. One does not have to be gay to be put into that certain position in which one just cannot go on with the old language game any further, where the facts 'buck' hard enough to throw one 'out of the saddle'. Think small: think of not being allowed a haircut or a drink of water from a certain fountain because of one's skin colour, think of not being able to enter the library or just walking outside because of the femaleness of your genitalia. Think of not being taken seriously because the accent was not quite right. Think of not being served bread and butter because – well, because you were in drag. Think of the dogs that the police let fly against your body because of a simple outfit, or rather, a not-so-simple outfit, or rather a simple outfit worn against a not-so-simple rule, a proposition rather than a directive *per se* that set the routing just as effectively as if there were an AK47 at your head. To wit: 'Never stray from the path; and if you do, you will pay for it.'

Twenty years ago, the NY Stonewall riots occurred over rolls and butter, dogs and lipstick, skirts and trousers covering the wrong behinds. It was a marker, a nodal point, and as such, it was not exactly the 'beginning' of the movement, but it certainly was a first. Indeed, there was no 'beginning', just copies, repetitions, repetitions of the origins – molecular repetitions of the truths, which never existed except as that which coalesced into a multiplying and endless return to the 'that which lies around it', a return which in itself was not, necessarily, 'of', 'alongside' or 'within' the same 'space'. Rather, it was one that emerged exactly in the realm of an 'excluded middle', though not as 'neat and clean' as will be narrated here, and certainly sans any sense of chronological ordering. There were twists and turns and belly-aches throughout.

For people whose sexual hunger was best fed through a diet of same sex products, the tearing away against the 'sureness' of the game, at least on the face of it seemed quite simple. An 'us' was neatly constructed, an 'us' against 'them'. The 'us' were all those people who were interested in having same-sex sexual relations, but in so doing were (and in many quarters, still are today) subjected to a variety of difficulties ranging from prison sentences to job losses to 'queer bashing' to familiar disownership; and sometimes facing all this and more. The 'them' were all the rest. We had before us a 'p'/'not-p'. Given this 'us–them' formulation, the task was obvious: organize ourselves, all the 'us's', i.e. all the lesbians and homosexuals, to fight against those pathetic injustices established both at the level of commonsense and the law; indeed, established precisely in terms of a commonsense *as* the Law.

On the one hand, it has to be said, that that strategy, though not fully articulated and certainly without actual appointed 'leaders' *per se* (though organic intellectuals seemed to pop up everywhere: newspapers, plays, the music 'scene', and even in the hallowed halls of academia, though kitchens and laundramats not to mention the alleyways proved a far more likely breeding ground); that that strategy around claiming an identity 'against' the 'them', a strategy we called 'gay' liberation, yielded up something rather extraordinary: a more or less *public* space yawning before us, and indeed, engulfing us with the promise of a future. With spurts of pleasure erupting everywhere (or so it seemed); that is, with the transgressing of a status quo pleasure – no, more than that, with the utter transgressing of a body-flesh in the name of Righteousness itself, this movement created a raw and very fluid public space, one that was both visible and invisible – in the streets and not – at precisely the same moment. God *was* a whore. A whoring God slashing at the image of an 'infinite goodness' with the deep and unforgiving crop of Sadean corruption, i.e., with the violent cut, the wicked mark – a (re)marking of the deep slash which had originally culminated precisely into that which was its preface, i.e., that 'God is dead'.

But unlike the Nietzschean transgression whereby the death of God heralded what he called the 'empty purity of its transgression',[13] this sexualizing play, this play which tossed God beyond the limit, corrupting the not-corruptible, produced a transgression which could never remain 'pure', could never remain 'innocent'. Indeed, by admitting – in the fullest sense of that term, i.e., confessing and allowing – God into a body whose sexual appetites were to be held out as unquenchable and available to all, for a price, the proposition, 'God is a whore' at once engulfed the promise of its own future on an infinitely vanishing ground. 'In history as in nature,' the 'Old Mole' was wont to have said, 'decay is the laboratory of life.'[14] To this Bataille simply added the requisite dirt.

On the other hand, a peculiar thing also began to emerge, particularly in terms of those involved with the movement itself: it became dauntingly obvious that this entity called a homosexual or a lesbian, that is to say the 'gay' male or female, could not quite be traced, no matter how transgressive of the Law she or he became. Because the identity that had been marked out against an always-already given 'that' – i.e., as Not Heterosexual; or Not Vaginal Penetration (for the 'real' lesbian); or Not Macho (for the queens); or Not Feminine (for the butches); or Not Monogamous for the men; or Not Perverse, for the women – no longer sufficed. Where, oh where, for example, did the gay transsexuals fit in?

Indeed, the very fact of transsexualism (not to mention, at least in earlier times, the fact of the hermaphroditic body)[15] foregrounded this other complication: organizing around an 'us' had been predicated on an assumption that the sexual was itself 'natural', and that, as such, sexual orientation (in whatever direction one might go) did not involve 'taking a decision'. This certainly became problematized by those who took the decision to alter their genitalia and secondary sexual characteristics – and then had the 'temerity' to rename themselves 'gay' if investing in same sex sexual relations on the basis of their post-surgical, and hence not 'naturally' given, sexual organ status.[16]

And while we're on the subject of 'what is natural', a whole new series of complications burst forth, particularly for the lesbian, though not exclusively. For, if one were 'straying from the path', and if the path she were straying from was organized around the body, and more particularly, the sexual body, the possibility existed that this lesbian might conceivably be engaged in (or hoping to be engaged in) very similar sexual practices that in days of yore were located – or at least presented 'as if' they were located – in the exclusive domain of the masculine. For example, her choosing to be among multiple sex partners engaging in a group bondage and whipping situation; though sometimes the scenario might include transsexuals (who might or might not be gay); or it might include gay men who might be 'clones' or 'queens' or whatever; sometimes the situation could emerge in terms of exchanging roles; or sometimes it might be a romp with two members of the same 'role', and so on, and so on, and so forth. Would *this* woman who had strayed from the path in such a manner, still be having a thing called lesbian-sex, if it included all these other referents? And what if she did most of these things half the time, but had a monogamous relationship (or none) during the other half? Should she be considered only half a lesbian? Which half?

These practices, emerging in the dim light of a semi-public, semi-underground, semi-private, and oft times fictional locality, could never be accounted for, let alone addressed, in terms of the panoptic gaze of an 'either/or'. For when we put the 'flickering picture' in those terms, we would never be able to catch sight of her; only a cheap and fleeting imitation. For her 'technique' or practices, were (and remain) themselves constitutive of, and determined by, the field of a subversive doubling – the de-(de-negation) – a kind of catachresis, as Derrida would put it, that embodies a *play* 'at and with' the return of 'a' truth, itself subject to the play of the 'game'.

This lesbian/gay pervert is the 'not'-'not' of the Other. She/he embodies and regurgitates the space of a whoring God; and yet, in so doing, replaces the binarism of a (sutured/closed notion of) Law/transgression with the corrupt and sublime 'nodal point' of pleasure. It's a kind of masochistic laughter-pleasure-play: one which both submits to a Law (which itself can never be fixed); and at the same time erupts somewhere *past* the sacred and the profane.

11c DEMOCRATIC CELLS, MEMORY AND THE QUESTION OF THE URBAN-CITY

Does it now seem so odd that this kind of being may quite possibly be one of, if not the, 'expression' of a radical pluralistic environ, one that disrupts in her multi-sexualized, and/or multi-gendered, wake the very notion of community? Or to put it slightly differently, is it so hard now to envision that the kind of democratic possibilities a quantum philosophy/or space-time logic suggests has more to do with 'impossible socialities', what could best be understood as quasi-social-somethings (or quasi-social-nothings) or in a word, 'cities'?

For I am speaking here of the kind of entities that no longer facilitate binarist community politics *per se* (and with it, 'alliance' building, replete with homogeneous notions of whatever movement they are with or against). Rather than excavate a sense of identity politics based on the Law and its transgression, the emphasis is both on the kind of city-scape, whose very reproduction is based on, maintained by, and emits of, an irreducible instability of discrete social elements; and at the same time, posits the 'communality' of those social elements as themselves, 'city', as themselves, fractured entities, fractured 'scenes', be they 'leather' or 'drag' or 's&m' or 'vanilla' or all of the above or none. For the whole focus now comes upon the importance – no, the *necessity* – to re-cover 'urban-ness' in all its anomie, and rather chaotic, heterogeneity, if we are indeed serious about creating a radically pluralistic and democratic society.

This is not for a moment to champion the 'individual' over the social (or for that matter, the reverse). This is to rethink the entirety of the paradigm, once we have shifted into fourth-dimensional gear – once, that is to say, we deal with the fact of being in (and buffeted by) the information age. It would seem to me that urbanity, and the cityscapes to which it points, is precisely the antidote to preserve or maybe invent, against a crushing and totalized 'community' (already being established on all sides of the political spectrum), what it is to stand before the Law, and to do more than just 'survive'.

A final point. If it no longer makes sense, on the whole, to organize 'against' the x *per se* – or to put it slightly differently – if we are now entering the epochal space of the 'not-space', where all infinity is corruptible *and* perverse, how does the political find its creative and wild possibility (and in that sense, its 'affirmation')?

One way – not *the* way, necessarily, but one way – is to take on the task of 're-re-membering'. But here I am speaking of a peculiar refurbishing (and/or) re-covering of memory, one having nothing to do with the reclamation of some long ago past. There is no interest here to go back in time to search for the future, no attempt to recoup an 'origin' or 'rationality' or 'continuity' to that which makes the present tense present. There is no attempt to resurrect Experience on the back of our sapphic whore and make the memory itself into a nodal point.

This is a call instead for a living memory: a popular memory, as Foucault had for so long insisted; a mourningful memory as Derrida so well set out via his homage to Paul de Man, one which *re*-covers the possible impossibility, the play of the game, over and over again. One which takes as its leverage an 'as if', i.e., one which re-members the 'that which lies around' in the spatiality of an excluded middle, and in so doing, takes it out of its impossible burial site or impossible void and *creates* it, tastes it, chews it, tries it on 'for size'. It is a conscious re-covering, healing, and narration of possibility. It is a mapping or envisioning of the 'that' of what we might become, based not on idle day-dream or fancy or even 'experience' *per se*, but rather, rooted precisely in *real* life, in all its precarious and profane ways. Without it, we are at the mercy of any other game; indeed, of many other games whose nightmares are not of our own choosing. But with it, we – this non-homogeneous, diverse and fluid 'we', politically cemented, hegemonically inscribed – this 'we' has the chance, the possibility (though only a possibility), to set an agenda, to be in a position to take a decision. It's an impossible mapping, an envisioning of the 'that' and the 'what' of the democratic possibility in a so-called 'postmodern' world. A kind of practical romanticism rooted in an 'as if' and fractured by the unequivocal perversity of the body itself.

After all this, maybe it's Nietzsche who caught it best:

> 37 You run on *ahead*? – Do you do so as a herdsman? or as an exception? A third possibility would be as a deserter . . . *First* question of conscience.
>
> 38 Are you *genuine*? or only an actor? A representation or that itself which is represented? – Finally you are no more than an imitation of an actor . . . *Second* question of conscience.
>
> 39 *The disappointed man speaks.* – I sought great human beings, I never found anything but the apes of their ideal.
>
> 40 Are you one who looks on? or who sets to work? or who looks away, turns aside . . . *Third* question of conscience.
>
> 41 [But] Do you want to accompany? or go on ahead? or go off alone? . . . One must know *what* one wants and *that* one wants. – *Fourth* question of conscience.[17]

NOTES

1 Paper delivered at The Institute of British Geographers Annual Conference, 7–10 January, 1992, Department of Geography, University College of Swansea, Wales. An earlier draft of the initial positions was presented at The Centre for Theoretical Studies, Essex University, March 1991.
2 F. Nietzsche, 'Twilight of the Idols', in *The Portable Nietzsche*, trans. W. Kaufmann, Middlesex, Penguin, 1976, p. 468.
3 Frank Cunningham, *Democratic Theory and Socialism*, Cambridge, Cambridge University Press, 1987, summary subsection: 'The empty space of democracy', p. 73. But see also Lefort's explicit discussion in, for example, his 'Human rights and the welfare state', in *Democracy and Political Theory*, trans. D. Macey, London, Polity Press, 1988, especially pp. 32ff.
4 Gaston Bachelard, *The Poetics of Space*, trans. Maria Jolas, Boston, Beacon Press, 1964, p. xviii.
5 Martin Heidegger, *Being and Time*, trans. J. Macquerie and E. Robinson, Oxford, Basil Blackwell, 1967 (7th edn), para. 23, p. 143.
6 For one of the most succinct and far-reaching assessments of the various problems inherent with an acceptance of some form of self-reflexivity as an anchor for truth, see for example, Rodolphe Gasché's well-known *The Tain of the Mirror: Derrida and the Philosophy of Reflection*, Cambridge, Mass., Harvard University Press, 1986, especially Part I, 'Toward the limits of reflection', pp. 13–108.
7 Phillipe Lacoue-Labarthe, *Topography: Mimesis, Philosophy, Politics*, with an introdiction by J. Derrida, Cambridge, Mass., Harvard University Press, 1989.
8. For a useful explanation of this point, see Stephen Hawking's *A Brief History of Time: From the Big Bang to Black Holes*, London, Bantam Press, 1988, pp. 23–37.
9 Ibid., p. 34.
10 L. Wittgenstein, *On Certainty*, trans. G. E. M. Anscombe, ed. G. H. von Wright, London, 1969, §125, p. 18e.
11 Part VI, *Mathematics*, §31, p. 333.
12 Part VII, *Mathematics*, §27, p. 395.
13 Nietzsche, in his *Gay Science*, framed the death like this: 'From the moment that Sade delivered its first words and marked out, in a single discourse, the boundaries of what suddenly became its kingdom, the language of sexuality has lifted us into the night where God is absent, and where all our actions are addressed to this absence in a profanation which at once identifies it, dissipates it, exhausts itself in it, and restores it to the empty purity of its transgression', F. Nietzsche, *The Gay Science*, p. 108, as quoted in M. Foucault, 'Preface to Transgression', *Language, Counter-Memory, Practice*, Oxford, Basil Blackwell, 1976, p. 31.
14 K. Marx, as quoted by G. Bataille, ' "The Old Mole" and prefix sur in the words surhomme and surrealist', *Excess of Vision*, p. 32.
15 One of the most interesting studies on the question of 'decision-taking' with regard to the hermaphroditic body is the well-known manuscript of her/his own life, by Herculine Barbine. In France, by a certain age, the hermaphrodite had to 'choose' to which sex they should forever be assigned. In the twentieth century this 'problem' has all but vanished, since sex re-assignment often happens at birth or before puberty. But see further, *Herculine Barbine*, with introduction by Michel Foucault, New York, Pantheon Books, 1980.
16 Certain feminists were really rather enraged over whether or not the transsexual woman who related sexually to other (biologically pre-surgical) women could be considered 'lesbian'. Some went as far as to accuse the male-to-female sex change as having undergone the operation purely to 'take over' the feminist movement.

See, for example, J. Raymond's outrageous *The Transsexual Empire: The Making of the She-Male*, Boston, Beacon Press, 1979.

17 F. Nietzsche, *Twilight of the Idols and The Anti-Christ*, trans. R. J. Hollingdale, Harmondsworth, Middlesex, Penguin, 1968, pp. 26–7.

REFERENCES

Bachelard, G. (1964) *The Poetics of Space*, Boston: Beacon Press.

Cunningham, F. (1987) *Democratic Theory and Socialism*, Cambridge: Cambridge University Press.

Foucault, M. (1977) *Language, Counter-Memory, Practice*, Oxford: Basil Blackwell.

—— (1980) *Herculine Barbine*, New York: Pantheon Books.

Gasché, R. (1986) *The Tain of the Mirror: Derrida and the Philosophy of Reflection*, Cambridge, Mass.: Harvard University Press.

Hawking, S. (1988) *A Brief History of Time: from the Big Bang to Black Holes*, London: Bantam Press.

Heidegger, M. (1967) *Being and Time*, Oxford: Basil Blackwell.

Lacoue-Labarthe, P. (1989) *Topography: Mimesis, Philosophy, Politics*, Cambridge, Mass.: Harvard University Press.

Lefort, C. (1988) *Democracy and Political Theory*, London: Polity Press.

Nietzsche, F. (1968) *Twilight of the Idols and the Anti-Christ*, Harmondsworth: Penguin.

—— (1976) 'Twilight of the idols', in *The Portable Nietzsche*, Harmondsworth: Penguin.

Raymond, J. (1979) *The Transsexual Empire: The Making of the She-male*, Boston: Beacon Press.

Wittgenstein, L. (1969) *On Certainty*, trans. G. E. M. Anscombe, ed. G. H. von Wright, London.

—— (1983) *Remarks on the Foundation of Mathematics*, trans. G. E. M. Anscombe, Cambridge, Mass.: MIT Press.

12

CONCLUSION:
Towards new radical geographies

Michael Keith and Steve Pile

We believe that this book demonstrates that all spatialities are political because they are the (covert) medium and (disguised) expression of asymmetrical relations of power. There can be no simple celebration or condemnation of transgression – the movement from one (political) place to another (political) place. Instead, there must be a commitment to the continual questioning of location, movement and direction – to challenging hegemonic constructions of place, of politics and of identity.

We would like to take this opportunity to make three small points concerning *Place and the Politics of Identity*. One relates to the relationship between ethical, epistemological and aesthetic geographies; the next to space, boundaries and closure; and, the last to spatiality, hybridity and radical politics. This is a different cut on arguments which have already been made, but we believe this raises issues which must be confronted. While many of these concerns have been raised before and although we do not claim to exhaust the discussion of these problems, we would like to suggest that taken together – and in the context of the book as a whole – there is here a movement towards new radical geographies.

ETHICAL, EPISTEMOLOGICAL AND AESTHETIC GEOGRAPHIES

As Harvey points out in chapter 3 of this volume, there is a need to use epistemology to identify the relationship between different differences. He argues that epistemology – if it is to fulfil this function – must be seen as a situated situated knowledge. Without retelling his argument, knowledges must be able to situate the knowledge that they claim to hold and they must be able to situate themselves. An analogy: they must be able to look through one eye while holding a mirror to the other to see itself looking. Not an impossible feat, but well beyond the capabilities of many so-called critical theories.

We would like to add a little more to this story: ethical and aesthetic geographies. Where epistemological geography would disrupt the distinction between inside and outside by being both the object and subject of the gaze (and therefore more than this), so too ethical and aesthetic geographies would disrupt the hidden ethics and aesthetics of a discipline in love with its own purity and beauty but quite incapable of describing it. This is not quite true; geographers have periodically described the love affair they have with their art. The sexual analogy is significant. Geographers – detached and male – gaze upon the landscape which is filled with seductive pleasures (see Rose, forthcoming). Landscape is Woman. The ethical and aesthetic dimensions are naturalized; the power of the geographers to gaze on the beauty of the world is unquestioned because it is associated with the power of men to stare at women, who are thus objectified.

There are intimate connections between epistemology, ethics and aesthetics in the construction of geography. Once J. K. Wright said:

> A great deal had been written and more said about the nature of geography; far less about the nature of geographers. Could we subject a few representative colleagues to a geographical psychoanalysis, I feel sure that it would often disclose the geographical libido as consisting fully as much in aesthetic sensitivity to the impressions of mountains, desert, or city as in an intellectual desire to solve objectively the problems that such environments present.
>
> (Wright 1947: 9)

A geographical libido? Solving problems? Aesthetic sensitivity? Such themes recur; the double fascination and horror with landscapes of beauty and ugliness is a perennial theme in the work of humanists, Marxists and logical positivists (whether human or physical geographers) *ad nauseam*. Yet situated knowledges neither threaten nor confirm this work, they only demand that the discursive conditions of production of analytical concepts be rendered explicit in terms of their historicity and their spatiality. Beyond this, all the old epistemological debates remain untouched, neither more nor less soluble than they ever have been.

Starting with notions of ethics, aesthetics and epistemology, once more returns us to a spatial metaphor: position. There is nothing necessarily frightening or necessarily relativist about this. Four-dimensionality and the contradictions of quantum physics and Newtonian physics may or may not be hypothetically reconciled in a theoretical world of 'n' dimensional space; but, as Massey pointedly remarks in Chapter 8, 'Newtonian physics is still perfectly adequate for building a bridge'.

In the end, progressive politics and new radical geographies surely must rest ultimately on the rejection of the absolute Kantian distinction between the realms of ethics, aesthetics and epistemologies. All articulations of spatiality invoke cognitive mapping, ethical markings and aestheticised territories. While recognizing – contingently, strategically – that each are sites of their own, sometimes different, tensions, they are only distinct if such values are abstracted, divorced from their social context.

MICHAEL KEITH AND STEVE PILE

SPACE, BOUNDARIES AND CLOSURE

In Part 1 of the introduction, we suggested that the vocabulary of space does not guarantee a radical project – and that this was as true in the case of politics in London's Docklands as in the politics of black diaspora. Space guarantees nothing – whether it is seen as hegemonic power, counter-hegemonic power, or something else entirely. We suggest that spatial mobilization is a double-edged sword.

Politics is invariably about closure; it is about the moments at which boundaries become, symbolically, Berlin Walls. These politics hermetically seal these boundaries, creating spaces of closure; on one side 'the goodies' and on the other 'the baddies'. An eternal struggle between Good and Evil; are you for us or against us? If you are not part of the solution, you must be a part of the problem. 'Us' versus 'them' logic is worked out in many arenas of political struggle; simplifying more than a little, feminism mobilized women against men; class politics, labour against capitalists; anti-imperialism, colonized against colonizer. The dichotomized social movements that resulted and all the combinatorial associations associated with them (e.g. socialist feminism) internalized an imagery that was as much symbolic as it was real; it had to assume the existence of the masses they hoped they could mobilize, and who would be the agents of history. In Western societies at least, the masses have failed to materialize, mobilize. Indeed, it may well be argued that they have de-materialized (into the middle classes) and de-mobbed (into the underclasses). The sword cuts both ways. There is another way to look at this issue.

The processing of history into myth is an essential element of popular mobilization because conflicts in one part of civil society may be isolated and yet still prompt a symbolic class unity, a class-consciousness rooted in empathic readings of other people's struggles. One community's struggle is tied to others not solely through compatibility and community of interest, but also through the meanings with which all forms of popular mobilization may be endowed. Constructions of 'resistance' tie together different struggles. This is the context in which black community support for the miners' strike was so outstanding (see Fine and Millar 1985; Gilroy 1987).

The point is that it is the ground on which these closures take place that must be central to any praxis which communicates between situated knowledges and momentarily, strategically completes the always incomplete process of identity formation. As we have stressed over and over again, this does not imply a slide into relativism, it instead points the way towards those discourses that can deploy spatialization in the communication between difference, such as in certain forms of postmodern jurisprudence.

What we are arguing for is a limited autonomy between different kinds of spatiality. Moreover, no metaphors of space can be confined entirely to an ethical, epistemological or aesthetic realm, they are inevitably stained, by accident or design, with both hegemonic readings and claims to truth. But they are never exhausted by them. So when Richard Wright suggested famously that black and white writers are forever in contest over nature of reality, there is a very real sense in which we can acknowledge the value of such an insight without a necessary resort to ethnic essentialism.

SPATIALITY, HYBRIDITY AND RADICAL POLITICS

In part, we are trying to locate new radical geographies within a current moment of political theorization that takes spatiality as central, even if almost unwittingly so. To be honest, we believe it is a moment of credentializing opportunity as well as a moment of political relevance. This is a moment in which spatiality appears about to take its place alongside historicity as one of the key terms in theories of conjuncture and situated knowledges. Yet the academy clearly still neglects the problematic nature of spatiality despite 'space' becoming increasingly central to social and cultural studies.

To take one last case of how contemporary theorization so often simultaneously draws on and confuses notions of spatiality, we will turn to an article by Homi Bhabha (1992). The article draws much on Lacan's understandings of identity (via Laclau) and like us takes the *Satanic Verses* affair as one of its reference points. Bhabha, like Rushdie, celebrates hybridity and syncretism, he also privileges a time-lag which is defined in the following terms: 'The time-lag that I want to inscribe for the analysis of postcolonial discourse as a productive hybrid "betweenness", relocation and reinscription' (Bhabha 1992). Here the locus of difference becomes tied to historicity rather than to spatiality. Bhabha's sense of a cultural void – which needs to be filled if cultural survival is to be achieved – is to be filled by a discursive remembering. Without detracting from the importance of remembering, he could equally plausibly talk of cultural in-filling. Not so much a time-lag, as seeing what is there: making the disappeared visible.

As significantly, this collection suggests that the spatial references in contemporary social and cultural theory are extremely complex: we cannot simply privilege particular spatial locations. Instead we have to unpack the specific spatialities that these geographical vocabularies evoke. The invocations of spatiality so common to contemporary cultural theory have much to offer geography, but equally the understandings of the complexity of spatiality itself is a long-standing subject of geographical interest. In part, cultural theory makes geography's familiar appear strange, just as geography can and should prompt cultural studies to reflect on its common sense spatialized vocabulary. In a small way this book attempts to open this dialogue.

Typically, it is never going to be unproblematic to use spatial metaphors such as centre/margin or inside/outside because of the danger of privileging or even romanticizing one side of the coin (whether by the powerful or the powerless – itself a problematic dualism). For example, Homi Bhabha argues that 'it is from the affective experience of social marginality that we must conceive a political strategy of empowerment and articulation' (Bhabha 1992: 56).

The importance of experience cannot be denied, but it relies on a strong sense of who you are and where you are: it relies on a 'subject position', but this has to be set against Diana Fuss's warning that subject positions cannot be assumed because they are themselves social constructions (Fuss 1989: 29–36). All sorts of notions can be carelessly connoted by the unproblematic use of spatialized language. Speaking positions, viewing positions, embodiment, conjuncture, context, iconic references, symbolic rallying points, and imagined geographies can be both confusingly conflated in conscious or accidental assertions of the salience of the spatial. Yet this volume demonstrates that this slippage is unnecessary. Spatialities do subsume contradiction and confusion, or at least attempt to, but – because they may also fail, especially when placed side by side – they open up the spaces of political resistance and progressive politics.

. . . AND FINALLY

We would like to end by making two last points about *Place and the Politics of Identity*, and these points relate to spatiality. For us, the term takes on a central theoretical location – it becomes the name for the surfaces of articulation between regimes of power. We want to highlight two aspects of spatiality (though this again should not be seen as exhaustive of the term).

First, we would like to argue that spatiality needs to be seen as the modality through which contradictions are normalized, naturalized and neutralized. Politics is necessarily territorial but these territories are simultaneously real, imaginary and symbolic. We would like to illustrate this with a passage from David Widgery's book, which is an autobiographical account of being a doctor in the East End of London. He draws certain conclusions from the human misery that supposedly *laissez-faire* capitalism produces. Widgery is caught between sympathy and anger; between caring for people who live day-in, day-out with deprivation and neglect and knowing that those same people can be racist bigots and have mugged him for a packet of pills. Nevertheless Widgery believes that

> there is a palpable set of East End values: that you shouldn't jump the queue, that you are not just in it for yourself, that it's right to help people in trouble and that loyalties matter more than cash. Together they mean a solidarity, if not of factory or union branch, then of everyday life. Often they are expressed most strongly where children are concerned. But these beliefs are under attack every day by a highly coherent commercial culture which has a price for everything but knows nothing's value. They inevitably break down into a muddle of incentives and optings-out and couldn't-care-lesses. It rains demands and the streets are awash with despair. So why not do as culture says? Do your

> own thing rather than do the right thing. And then wonder why you can no longer assemble anything from the bits and pieces of identity you can salvage. And discover you have nothing to give your children that works without batteries.
>
> (Widgery 1991: 74)

The East End is a hard place to be poor, sick and old. And the East End has come to stand as a physical expression of London's poverty, sickness and ageing. People and place locked together – the only escape is for those able to. Those who are left are locked into a vicious circle of declining opportunities and disinvestment. There are many conflicts in the East End; there is a deep-seated racism and the ever-present possibility of violence. This cannot be ignored or denied. Yet, people continually mobilize around the place, its politics and their identity.

It follows that, second, spatiality should simultaneously express people's experiences of, for example, displacement (a feeling of being out of place), dislocation (relating to alienation) and fragmentation (the jarring multiple identities). Spatialities represent both the spaces between multiple identities and the contradictions within identities. There are gaps not simply between the identities we present in different socio-spatial settings, but also within particular identities. In this way, to say that someone is a white middle-class university lecturer and that this person shaves and wears spectacles, would not be enough to specify that person. It is not enough to tell what their political or sexual preferences are, or even to know their gender – even while we might believe (probably accurately) that this academic is also male, married with 2.4 children, middle-aged, has a Volvo car and is hardly a radical. Even this would not exhaust a person's identities, nor the contradictions between them or within them.

In these spaces, there is hope for us all; the deconstructive and radical move is to refuse to let things – in this case, particular political mobilizations around particular concepts of space – settle. Space is constitutive of the social; spatiality is constitutive of the personal and the political; new radical geographies must demystify the manner in which oppressions are naturalized through concepts of space and spatialities and recover progressive articulations of place and the politics of identity.

REFERENCES

Bhabha, H. (1992) 'Postcolonial authority and postmodern guilt', in L. Grossberg, C. Nelson and P. Treichler (eds) *Cultural Studies*, London: Routledge, 56–66.

Fuss, D. (1989) *Essentially Speaking: Feminism, Nature and Difference*, New York: Routledge.

Fine, B. and Millar, R. (1985) *Policing the Miners' Strike*, London: Lawrence & Wishart.

Gilroy, P. (1987) *There Ain't No Black in the Union Jack*, London: Hutchinson.

Rose, G. (forthcoming) *Feminism and Geography*, Cambridge: Polity Press.

Widgery, D. (1991) *Some Lives: a GP's East End*, London: Sinclair-Stevenson.
Wright, J. K. (1947) '*Terrae Incognitae*: the place of the imagination in geography', *Annals of the Association of American Geographers* 37(1): 1–15.

INDEX

abuelas de Plaza de Mayo (grandmothers of the Plaza de Mayo, Argentina) 108
Adams, Mary Louise 93
Adorno, Theodore 9
aesthetics 26, 32, 220–1
affirmative action (USA) 48
Alcoff, Linda 94
Alfonsín, Raul 111
alliances, progressive political *see* politics
Althusser, Louis 71–2
Alvarez, Sonia 103
amorality 136
anti-essentialism 94, 99; essentialism of anti-essentialism 98
anti-humanism 95, *see also* dualisms
Argentina 102–13; anti-women measures 105; gender and nationalism 104–5; guerilla movements 106; military government 105–6; women's legal status in Argentina 104
author, power of 122
authority 118, 122, 137; authority to speak 123; black authority 18; moral authority 57; post-colonial authority 16–20
autobiography 121

Bachelard, Gaston 208
Balibar, Etienne 178
Barnes, Trevor 69
Bartky, Sandra Lee 92
Being and Time 208
Being-in-the-world 208
Benjamin, Walter 8–9
Berlin (Germany) 8
Berman, Marshall 7, 89
Bhabha, Homi 19, 32, 33, 141, 158, 223–4
binary opposites/oppositions/ordering *see* dualisms
biography 131–2
Birmingham (UK) 164, 173, 177
black 17–18, 165–71, 176, 183; black Britain 164, 168; black communities 17, 18, 118, 167–8, 170, 177; black cultural forms 167, 170; black diaspora 17, 170, 172, 222; black discourses 168; black identity 18, 166–7, 177; black intellectual forms 17; black journeying 176–80; black men 165; black migrants 165; black nationalism 186; black politics 164, 165, 168–9, 170, 172; black presence 166, 170; black radicalism 165, 190; black representation 167; black settlement in Britain 163–5, 168, 172, 177, 179; black social movements 165, 168
body, the 193, 213–15, 217; the flesh 212
Bondi, Liz 111
borderlands 78
boundaries 222–3
Bourne, Jenny 84, 94
Bristol (UK) 162, 164, 165, 173, 177
Broiler Belt, the (USA) 41
Brunt, Rosalind 84
Buenos Aires (Argentina) 107, 108
Butler, Judith 92

Canary Wharf (London, UK) 11, 13, 16
Capital, Das 42
capitalism 7, 24–5, 79, 89, 167, 193
Cardiff (Wales) 163, 164, 165, 173
Catholicism 104; Catholic faith and

feminism 109; Catholic nationalism 111; Catholic notions of purity 104
childhood 128, 133
citizenship 108, 122
city, the 7, 8–9, 21, 25, 26, 33, 118–19, 125, 131, 149, 164, 177, 180, 192–3, 207, 211, 216–17; cities and black cultural forms 173; inner cities 43, 45, 120, 124; race riots 164; urban deprivation 118–19; urban meanings 168; urban racial antagonisms 163, 172; urban regeneration 11; urban spatiality 193; urban-ness 207; women in cities 149
class 43, 89, 119, 168, 185; authentic working class identity 89; capitalist class interests 45, 47; and the Republican party (USA) 46; class consciousness 91, 222; class identity 91, 119; class oppression *see* oppression; class politics 44; class politics of knowledge 72; class relations 52, 60–2, 85; class struggle *see* struggle; class war 61; dominant class 46; ruling class 49; working class communities 125; working class consciousness 89–90; working class, the 18, 186
Clifford, James 78–9, 178–9
closure 30, 222–3; arbitrary closure 18, 28
cognitive mapping *see* mapping
commodification 89
Communist Manifesto, The 89
community 55, 90, 92, 103, 108, 110, 118, 119–21, 125–7, 128–37, 168, 177, 193, 207, 216, *see also* black *and* class *and* identity *and* white; communities of resistance *see* resistance; community identity 16; imagined community 171
Condition of Postmodernity, The 59
conjuncture 27
consciousness, 'true' 89
consciousness-raising groups 91, 95, 97
contingency 206, 211; theorizing contingency 26–31
cultural fusion *see* syncretism

Darby, Harold 157
Davis, Mike 156
de Beauvoir, Simone 151
de Bonafini, Hebe 102, 106, 110
de Certeau, Michel 16, 25, 130
de Man, Paul 217
democracy 112, 207, 213, 216–17
Derby (UK) 132
Derrida, Jacques 54, 174, 176, 217
Deutsche, Roslyn 25, 193
diaspora 17, 18, 188, 166, 172, 175–6; (cultural) politics of diaspora 16–20; diaspora and imagined geography 17, 19; diasporic boundaries 18; immigration 168; migrant identity 130; radical dispersal 169
dichotomy *see* dualisms
difference 16, 34, 55, 60–2, 63–4, 77, 124, 178, 184–9, 191–2, 194, 197; difference and universality 55; difference and working class rhetoric 52; group difference 57; politics of difference *see* politics; pure difference 30; racialized difference 17; socio-spatial differentiation 185
disappeared, the (Argentina) 102, 106–9, 112
dislocation 21, 27, 29, 144, 149, 154, 169, 173–6, 208, 210, 225
Disney World (USA) 9
displacement 78
domination. social 70; social domination and space 76
dualisms 7, 23, 26, 35, 36, 147–9, 151, 169, 175, 185–8, 189, 193, 195–6, 198, 209–10, 216, 222; capital/labour 147, 185; centre/margin 78, 191, 194, 223; colonizer/colonized 186, 191, 195, 222; global/local 7, 155, 195; good/evil 23, 210, 222; humanism/anti-humanism 84, 85, 91, 94, 96; inside/outside 18, 71, 109, 162, 173–6, 194, 209, 220, 223; male/female 136, 147, 186, 193; masculinity/femininity 36, 150, 175; master/slave 58, 169; mind/body 36; myth/reality 26; oppressor/oppressed 18, 58; personal/political 112; powerful/powerless 23; presence/absence 71; public/private 7, 102, 103, 108–9, 112, 136, 177, 193, 212; same/other 18, 169, and self/other 17, 63, 123, 186 *see also* selfhood *and* otherness; space/time *see* space/time; transcoding of value between dualisms 36, 112, 148–9,

175–6; us/them 35, 222; visible/ invisible 71, 109
Duncan, James 69

E = MC^2 210
Eagleton, Terry 150
Eco, Umberto 26
ego-psychology 87, 89
Einstein, Albert 210–11
emancipation 94, 96, 197
Emergence of Social Space, The 150
empowerment 62, 92, 224
Engels, Freidrich 50
epistemology 19, 51, 60, 62, 63, 72–3, 95, 190, 200, 220–1, *see also* dualisms; ambivalent epistemologies 31–4; politics of knowledge *see* politics; situated knowledge 19; situating situated knowledges 57–60, 220
equality 48; *see also* difference
essentialism 29, 58, 85, 92–3, 95–6, 97–9, 109, 111, 113, 150, 200; ethnic essentialism 19, 223; female essences 113; spatial essentialism 79
estrangement 133–4
ethics 26, 32, 206, 211, 220–1
ethnicity, new 32
ethnoscapes 176; global ethnoscapes 177; local ethnoscapes 178
everyday life 127, 137
excluded middle 209, 217
experience 95
exploitation 43–4, 56, 84

family 102, 104, 106, 108, 110–12, 127, 131
Fanon, Franz 19, 30, 31
Feijoó María del Carmen 110–11
female body, the 76, 213
female domestic sphere 110
female labour 105, 110–11
femininity *see* sexual identity
feminism 60, 87–8, 91–7, 105, 186, 195, 197; feminism and identity politics 91–7; feminism and socialism 91; feminism and the Catholic faith 109; feminist identity 91–2; feminist politics 84, 92, 95, 105; feminist theory 103–4, 192–3; hyphenated feminisms 93, 95–6; Latin American women's movement 103; multiple feminist identities 93; sisterhood 93; women's resistance movement 110
flâneur, the 8, 25, 33
Foucault, Michel 4, 54, 72–4, 76, 141, 154, 162–3, 197
fragmentation 186–8, 193, 216, 225
Fraser, Nancy 193
Freud, Sigmund 74, 84, 85–9, 97, 149; Freud and female sexuality 87; Freud's anti-humanism 88, 91
Friedman, Jonathan 7
Fuss, Diana 98, 111, 194, 224

Gallop, Jane 37, 96
gauchos, the (Argentina) 104–5
gay *see* sexual identity *and* sexual, the *and* sexuality
gaze, the 31, 118
gender 84, 87, 104, 146–51, 185; gender identities 104, 109; gender politics 104, 106, 108–9, 113, *see also* madres de Plaza de Mayo; gender relations 102, 104, 112, 120; gendering of space 192, *see also* dualisms; geography of gender 109
geography 73, 74, 79, 146, 159, 166, 184, 196–200, 208, 210, 221, 223, *see also* space; geographical description 157, 159, 199; human geography 155, 191, 197; imagined geography 17, 19, 195, 208, 224; political geography of the female body *see* female body; radical geography 34, 145–6, 153, 197, 206, 208–9, 211, 220–5; real and imagined geography 36, 195, 198–9
Gilroy, Paul 17–20, 33, 167
Glick, Tom 74
Glissant, Edouard 169
globality 166, 177, *see also* dualisms
God 119, 134–5, 209, 211, 214; God as a spatial metaphor 134, 216
Gramsci, Antonio 210
Greenham Common (UK) 38

Hall, Stuart 163
Hamlet, North Carolina (USA) 41–7, 58; *see also* Imperial Foods
Haraway, Donna 30, 31, 58, 60, 93, 99, 193
Hardy, Thomas 9
Hartsock, Nancy 58, 59, 61

Harvey, David 24, 30, 68
health and safety legislation, North Carolina (USA) 42, 45; *see also* Occupational Safety and Health Administration
Hebdige, Dick 196–7
Heidegger, Martin 208
Heller, Agnes 121
Hesse, Barnor 17
historicism 195, 200
historicity 27, 29, 221; radical historicity 29
history 146, 159, 166, 169, 199, 200
homosexuality *see* sexuality
hooks, bell 5, 33, 122, 179, 183, 184, 189–92, 197, 199, 200
humanism 135; liberal humanism 86, 88, 93–5, 185, *see also* dualisms
hybridity 20, 32, 220, 223–4
hybridization 20
hyperreality 26

identity 3, 6, 19, 26, 28–30, 58–9, 61, 63–4, 69, 76, 78–9, 84–5, 86–7, 90, 96–7, 103, 104, 113, 118, 119–20, 125, 128–37, 158, 162, 169, 172, 174, 176–7, 187, 192, 214, 225, *see also* black *and* class *and* community *and* diaspora *and* gender *and* sexual identity *and* white *and* women; collective identities 111; denied identity 106; fractured identities 91, 96; group identity 177; identity formation 16, 27–9, 31, 87–91, 147, 166; incomplete identities 31–4; identity and narrative 16, 162–4; local identity 177; marginal identities 194; maternal identity 109; multiple identities *see* multiple identities; natural identity 111; official identity 106; personal identity 129; political identity 64; politics of identity *see* politics; identity politics of place 20; post-colonial identities 19; shared identity 131; singular identity 35, 55; socio–spatial identity 9; travelling identities 78
identity politics 17, 31, 32–3, 64, 77, 84, 85–7, 92, 94–6, 97–8, 107, 113, 185–8, 192, 198, 216, *see also* black *and* class *and* feminism *and* sexual, the *and* white

immigration *see* diaspora
Imperial Foods chicken processing plant 41–2; (plant fire) industrial accident 41, 59, 60, 65, and lack of political response 44, 52
imperialism 196; cultural imperialism 56
incommensurability 17, 19, 23
India 132
individual, the 85–7
industrial organization 43
industrial reserve army 43
inequality, social 63; *see also* equality
injustice 51, 53, 56, 63, 167; *see also* social justice
inscription, surfaces of 27, 29
Islam 22, 28

Jackson, J. B. 7
Jacobs, Jane 7
Jameson, Fredric 2–3, 4, 5, 26, 68, 141, 142, 148–51, 155, 157–9
Jaquette, Jane 103
Jardine, Alice 175–6
Jay, Nancy 147–8
journeying *see* travel *and* black
justice 51, 62; *see also* social justice

King, Rodney, beating of 44
Kirkwood, Julieta 111
knowledge *see* epistemology; pure 30
Kovel, Joel 85
Kruger, Barbara 9

labour control 43
Lacan, Jacques 89, 91; Lacan and subjectivity 94
Laclau, Ernesto 27–31, 34, 141–6, 147–9, 151–4, 157, 169, 172, 173–5
Lake, Carol 117–19, 123, 126, 128, 131, 132, 134–7
language 122–5
language games 49, 57; *see also* Wittgenstein
Lash, Scott 7
Latin America 102; Latin American Catholicism 104; Latin American conservative nationalism 104; Latin American society 104; Latin American women's movement 103, 111
Law, the 212–16
Lefebvre, Henri 3, 4, 24–5, 26, 30, 36,

68, 76, 80, 142–3, 146, 197–8
Leigh, Mike 126
Levinas, Emmanuel 124, 125
Liverpool (UK) 162, 163, 164, 165, 173, 177
Lloyd, Genevieve 147, 149
local, breaking out of the 54–7
locality 120, 137
location 26, 69, 76–7, 78–9, 85, 98, 141, 156, 166, 189, 194, *see also* politics of location
locations of struggle 5, 34–5
London (UK) 26, 120, 164, 165, 173, 177
London's docklands (UK) 11–16, 222; Cherry Gardens 15; London Docklands Development Corporation 11, 13, 14–15; London's docklands as a symbol 14; Newham Docklands Forum 15; People's Armada 15; People's Charter 15; People's Plan for the Royal Docks 15; political economy of redevelopment of docklands 11–14; resistance to development 14–16; Rotherhithe Community Planning Centre 15; Royal docks 15; Surrey docks 15
London's East End (UK) 19, 224–5
Los Angeles (USA) 1, 25, 44, 156
Lyotard, Jean-Francois 54

madres de Plaza de Mayo (mothers of the Plaza de Mayo, Argentina) 38, 103, 107–13; madres de Plaza de Mayo and gender politics 108–9; community of identity 108
Making of the Black Working Class, The 164
male hegemony 124
male sphere 112
Man of Reason, The 147
man 149, 151, 212; man and mastery 212
man's world, in a 136
mapping 21, 69–70, 157, 193, 195, 199, 217; cognitive mapping 3, 68, 141, 221; impossible mapping 217; mapping and social control 70; social mapping of identities 77
Marable, Manning 166
margins 78, 136, 142, 189–91, 194; marginality 190, 191, 193–5, 197, 199, 200, 223; marginalization 56
Marx, Karl 42, 45, 50, 74, 84, 85–6, 89–91, 97; Marx's anti-humanism 90, 91
Marxism 53, 150, 186, 187, 195, 197
masculinity *see* sexual identity
masochism *see* sexuality
Massey, Doreen 18, 27, 29, 104, 120, 221
Meadowell estate (UK) 19
memory *see* remembering
men 149, 215
Menem, Carlos 111
metaphor(s) 68, 69–71, 74, 76, 79, 98, 103, 113, 141, 153, 173–4, 184, 189, *see also* spatial metaphors *and* dualisms; cartographic metaphors 70
migration *see* diaspora
military regime (Argentina) 107, 108, 112; military regime and civilian politics 109, 112
military repression (Argentina) 105
Minh-ha, Trin T. 194
mobilization 35, 53, 109, 113, 225; women's mobilization 102, 107
moral social order 126
morality 128, 136, 209
Morris, Meaghan 193
motherhood 110, 112, *see also* women
mothers of Plaza de Mayo *see* madres de Plaza de Mayo
Mouffe, Chantal 27, 172
Mouse, Mickey 9
multiple identities 49, 77–8, 93, 96, 127, 136, 225; multiple subjectivities 186

narratives 130, 163–4, 217; narratives of history 131
National Labour Relations Board (USA) 44, 46
nationalism, conservative *see* Latin America
negativity 27, 28
New Deal (USA) 44–5
New Reflections on the Revolution of our Time 27, 142, 143, 157
Nietzsche, Friedrich 207, 211, 217
Nuclear Free Zones (UK) 10, 25

object relations theory 89
Occupational Safety and Health

Administration (OSHA) (USA) 45–6, 47; *see also* health and safety legislation
oppositional politics *see* politics
oppression 1, 34, 43–4, 84, 85, 94, 96; class oppression 60; ecological oppression 56; five faces of oppression 56; multiple forms of oppression 57; oppressor's language 123; racialized oppression 17; sexual and racial oppression 64
otherness 50, 51, 58–9, 63–4, 125, 130, 171, 185, *see also* dualisms; multiple otherness 43; other(s) 52, 54, 58, 88, 162, 175, 190, 210, 216; unassimilated otherness 56, 57
outside *see* dualisms
outside, constitutive 27, 172

Paris (France) 8, 25, 151
Parmar, Pratibha 190
patriarchy 136, 193; capitalist patriarch 79
Péron, Eva 105
Péron, Juan 104, 105
physics 151–4; Newtonian physics 152, 154, 221; quantum physics 210, 221
place 23, 26, 97–9, 103, 119–20, 141, 225; place and history 104
Plato 49–50
Plaza de Mayo, madres de *see* madres de Plaza de Mayo
Plaza de Mayo, space of (Argentina) 110
pleasure 214, 216
Poetics of Space, The 208
political action *see* politics
political activists, agents, subjects *see* politics
political spaces 5, 37–8, 103, 105, 109–13, 144
politics 5, 109–10, 144, 146, 159, 177, 210, 216, 222; alliances, progressive political 36; class politics *see* class; collective politics 94; contested politics 36; cultural politics 184, 185, 187, 197, 199; feminist politics *see* feminism; gender politics *see* gender; modernist radical politics 187; oppositional politics 3, 84, 188; political action 3, 34, 62; political activists 103, 111; political mobilization *see* mobilization; political paralysis 53; politics and post-structuralism 94, 187–8; politics of articulation 30; politics of exclusion 57; politics of fulfilment 167; politics of identification 174; politics of 'race' *see* 'race'; politics of resistance *see* resistance; political subjects 27, 106, *see also* subjectivity, radical; politics of transfiguration 167; progressive politics 38, 47, 63, 211, 221; radical cultural politics 5; radical politics 34, 36, 84, 188, 220, 223–4; sexual politics *see* sexual, the; working class political organization 43; working class politics 47, 59
politics of difference 60–2, 183–4, 186, 189–96, 198, 200
politics of identity 5–6, 17, 30, 31, 34, 84, 91, 96, 97, 119, 186, 197, 225, *see also* identity politics *and* spatialization
politics of location 6, 76, 78, 142, 179, 189
politics of space 67, 142–3, 146, 154, 157, 162, 168, 224
polyvocality 125, 184
position/positionality 50, 57, 63, 69, 85, 98, 141, 187, 189, 208, 221; *see also* spaces *and* spatiality
post-structuralism 53–4
Postmodern Geographies 150
poultry industry, the (USA) 42
poverty, feminized 61, 64
power, 5, 29, 38, 73, 79, 95, 104, 112, 113, 120, 185, 210, 212–13, 220; landscapes of power 7; patriarchal power 198; power and space 76, 156; power and spatial metaphor *see* spatial metaphors; social power 63, 127
power-geometry 156
powerlessness 56
private, the *see* dualisms
Production of Space, The 24, 142
progressive politics *see* politics
psychoanalysis 87–8, 92
public, the *see* dualisms

quantum philosophy 206, 208–11, 216, *see also* physics

'race' 163, 168, 171, 178, 185, *see also* black *and* white; British racism 164;

British racist discourse 19; multiracial society 171; politics of 'race' 163, 165, 174, 177; 'race' and society 177; 'race' relations 19, 176–7; 'race' riots 164; racialization 171, 174, 177–80; racial antagonisms 171–6, 178; racial discourse 171; racial dislocations of modernity 169–70; racial harassment 171; racial politics 172; racialized capitalism 167; racism 23, 104, 167, 170–1, 178–9, 186, 193; racist riots 163
radical (cultural) politics *see* politics
radical contextualization 26–31, 31–4
radical geography *see* geography
radical openness *see* space
radical pluralism 207
radical undecidability 28
Rainbow coalition, the (USA) 44, 47
Ramdin, Ron 164
rationality *see* reason
reading 124–6
Reagan administration, the 46
reason 17, 19, 85, 124, 147–9, 212
relativism 49, 58, 77; cultural 19; *see also* spatial relativism
remembering 112, 130–1, 133, 207–8, 216–17; collective memory 120; men and remembering 131
resistance 1, 18, 33, 35, 38, 137, 169, 183, 185, 187–8, 191, 195, 197, 200, 222; collective identity of resistance 108; communities of resistance 5, 35–7, 103, 107, 110, 183, 194, 197; female resistance 107, 109; guerilla movements (Argentina) 106; new spaces of resistance 10, 200, *see also* madres de Plaza de Mayo; politics of resistance 54, 86; resistance to the redevelopment of London's docklands 14–16; sites of resistance 3
responsibility, ethical and moral 44; open-ended 53
Rich, Adrienne 76, 78, 79
Ricoeur, Paul 58–9, 121, 122
rights 52, 62, 94; human rights 37, 102, 111, 128; labour rights 46; women's rights 61, 105, 111; working class rights 61
Robbins, Bruce 78
Rosehill, Derby (UK) 117–19, 131, 134–5, 137
Rosehill: Portraits from a Midlands City 117, 118, 123, 125–8, 131
Ross, Kristin 150
Rowe, Dorothy 129
Rushdie affair, the 23
Rushdie, Salman 9, 20–1, 22–3, 31, 32, 158, 223

Said, Edward 128, 195–6
salvation 135–6
sameness 61–2
Satanic Verses, The 20–1, 22, 26, 31, 32, 33, 158, 223
Segal, Lynn 60–1
self(hood) 58–9, 64, 76, 85, 99, 125, 128–31, 136–7, 212; *see also* dualisms
sexual identity 87–8; femininity 87–8, 91, 110, 112, 148, 176; gay 96–7, 213, 215; masculinity 92, 104–5, 150; pervert 213, 216; queer 213
sexual, the 207, 215; gay liberation 97, 214; gay politics 213; queer bashing 214; politics 207, 211; sexual freedom 213; sexual politics 96–7, *see also* Stonewall riots; the erotic 213
sexuality 104, 185, 215; bisexuality 87; butch 215; female sexuality 87; gay 88; heterosexuality 88; homosexuality 88, 96, 97; lesbian 88, 215; mascochism 207, 211, 216; multiple sexualities 193; perversity 217; queen 215; sexual hunger 214; transsexualism 215
Shame 21
signification 124–5
Sillitoe, Alan 126
similarity 61–2
Sing, Carol 132, *see also* Carol Lake
sisterhood *see* identity politics
situated knowledge *see* epistemology
situatedness 50, 58, 60, 62, 63–4, 125; vulgar situatedness 57, 59
Smith, Neil 98
social being 95
social justice 34, 37–8, 44, 50–1, 56, 62, 63; social justice and postmodern criticism 50, 54, 62; postmodern death of justice 48–53; social justice and universalism 48, 53
social movements 53; new social movements 47
social networks 127

social order 128
social practices 129, 200
Soja, Ed 2, 4–5, 150, 156
solidarity 55, 61, 119, 133
Solomos, John 165
Sontag, Susan 9
space 2, 4, 5, 23–6, 36, 70–1, 74, 78–9, 80, 104, 110–12, 124–5, 137, 141–6, 151–3, 154–9, 169, 174–5, 177, 184, 189–93, 199, 207, 208–11, 213, 216, 217, 221, 222–3, 225; absolute space *see* space, absolute; abstract space 24, 104; actual space 5; cartographic space 70; deadness of space 79; death of space 74; depoliticized space 144; differential space 24, 25; fixity of space 5, 78, 143–5, 210; four-dimensional space/time 152; geographical space 73, 77, 78; geometrical space 2, 4, 143; ideological space 71; impossible space 37, 207; inclusive space 200; liminal space 7, 32; material (or real) space 70, 75, 191; mental space 4, 24; multiple spaces 102; not-space-space 207, 208–11, 217; opaque space 4; physical space 75, 144; political spaces, *see also* political spaces *and* politics of space; produced space 24, 73; public space 214, *see also* dualisms; real (or material) and metaphorical (or imagined) space *see* space, real and metaphorical; reconceptualization of space 6, 67, 189; relationality of space 74, 153; representation of space 4, 23–5, 32, 69, 75, 198; social space *see* space, social; space and political action 3, *see also* politics; space and the feminine 148–9; space as lack of politics 142–5, 157; space as routinized time 29; space as stasis 145; space in social and cultural theory 67, 196, 223; space of representation 23–4, 32, 33, 198; spaces of radical openness 5, 38, 189–92, 199; theoretical space 71, 124; third space 18, 192, 199; three-dimensional space 154; transitional space 7; transparent space 4; vernacular space 7; zero-dimensional space 69
space and time 67, 73, 104, 143–5, 147, 149–54, 156, 158, 162, 164–5, 168, 169, 171, 174, 180, 200, 211, *see also* space/time *and* time
space, absolute 25, 26, 74–6, 77, 79, 80, 153
space, real and metaphorical 5, 9, 35, 68, 73, 80, 103, 112, 130, 184, 185, 189–93, 197–9, 224, *see also* geography
space, social 4, 7, 24, 73, 74–5, 78, 80
space/time 36, 67, 73, 147, 150–4, 155, 158–9, 166, 175–6, 211–12, 216, *see also* space *and* time; multiplicities of space/time 158
spatial metaphors 1, 20, 35, 68–71, 72–3, 75, 77, 79, 80, 98–9, 103, 113, 133, 134, 191–2, 221, 222, 223, *see also* metaphors; power and spatial metaphors 76–9, 80
spatial, the 35–6, 143–5, 153, 155–7, 173, 199; spatial and the temporal 151–2, 155, 159; spatial concepts 67–8, 74, 79, 99; spatial discourse 71, 74–6, 184; spatial immanence 6, 7–8, 33; spatial language and social meaning 79; spatial logic of society 2, 68, 146; spatial modernism 79; spatial organization 146, 156; spatial politics 68, 69, *see also* political spaces *and* politics of space; spatial practices 71; spatial relativism 6, 8–11, 33; spatial terminology 141–3, 223; spatial vogue 2–5, 189–96
spatiality 4, 5, 6, 8–9, 16, 18, 24–5, 26–7, 29, 38, 73, 142–5, 153, 155–7, 173, 184, 189–93, 195, 197, 199, 217, 220–1, 222, 223–5; hybrid spatialities 23; impossible spatialities 38, 206, 210; metaphors of spatiality *see* spatial metaphors; multiple spatialities 19, 31–4; pure spatiality 158; spatiality and identity 9
spatialization 4, 18, 144, 148, 189, 195–6; spatialization of language 6, 7, 9, 16, 23, 32, 33, 71, 141–3, 184, 200; spatialization of political language 10; spatialization of politics 36
spatialization of the politics of identity 10, 20
Sphinx in the City, The 149
Spivak, Gayatri Chakravorty 59, 64, 74, 167, 194–5

Stacey, Margaret 126
Sternbach, Nancy 112
Stonehenge (UK) 38
Stonewall riots, New York (USA) 213
stories 129–30, 137
story-telling 118, 121, 123, 128–34; representing the experience of everyday life 119, 137; story-telling and space 124–5, 137; ways of story-telling 132
street, the 7, 8, 112, 128, 180
struggle 24, 62, 121, 180, 222; class struggle 52, 68, 185; labour struggles 44; *see also* locations of struggle; Carol Lake's struggle for identity 121
subject, the 84–5, 103; multiple subjects 187; subjects of history 112; subject positions 187, 224; the decentred subject 94
subjectivity, radical 36, 90, 183, 186, 190–2, 200
syncretism 18, 20, 33

Taussig, Michael 103, 113
Taylor, Charles 129
temporality 147–8, 155, 169, 173–4; black temporality 170
Thatcher, Margaret 137; Thatcher's Britain 118
Thomas, Clarence, supreme court nomination 44, 47
time 24, 73, 141–5, 149, 158, 174–5, 208, 210, *see also* space and time *and* space/time; cyclical time 144; times of reproduction 144
time and space *see* space and time *and* space/time
time-space *see* space/time
transgression 38, 207, 212, 213, 214, 215, 216, 220
travel 78–9, 176, 178–9; *see also* identity
trilectic, critical 200
truth 209, 222; truth-game 211

unconscious, the 85, 88
unemployment 43, 119
uneven development 185, 191
universality 48, 57, 60, 62, 63, 86; *see also* social justice *and* equality
Urban Development Corporations (UK) 13 *see also* London's docklands
urban, the *see* the city
utopia 134–6

violence 56, 112, 119, 124; violence and women 106, 110, 111
visible *see* dualisms

Wallman, Sandra 126
Weeks, Jeffrey 87, 88, 97
welfare state 61
West, Cornell 183, 184, 193
West, the 22, 28, 43, 70, 75, 147, 165–6, 175, 179, 195; western notions of reality 124, 131, 136
white 17, 77, 124, 171, 173; reactionary white 23; white communities 171; white identity 172, 178; white racism 169, 190; white society 171–3; white supremacism 166; white territorialists 174, 178; white working class 18
Widgery, David 224
Williams, Patricia 33, 34
Wilson, Elizabeth 88
Wilson, Elizabeth 149
Wittgenstein, Ludwig 49, 209, 212
woman 94, 97, 135, 149, 151, 176, 186, 215, 221; the 'woman effect' 176
women 87, 92, 102, 106, 112–13, 131, 135, 136, 149, 176; domestic and mothering role 105, 106, 109–12, *see also* motherhood *and* female labour; female subjects 104; women and violence 106, 110, 111; women's identities 102; women's legal status (Argentina) 104; women's lives 103, 111; women's mobilization *see* mobilization; women's movement *see* feminism; women's rights *see* rights; women's struggle for identity 121
working class politics *see* politics
working class rights *see* rights
working class *see* class
Wright, J. K. 221
Wright, Patrick 120–1
Wright, Richard 223
writing 124–125

Yearning: race, gender, and cultural politics 189–90, 192
Young, Iris Marion 34, 56, 58, 61, 92, 193

Zukin, Sharon 7–8, 27